ELECTRIC MACHINES AND TRANSFORMERS

ELECTRIC MACHINES AND TRANSFORMERS

Leonard R. Anderson

Southern Alberta Institute of Technology

RESTON PUBLISHING COMPANY, INC.
A Prentice-Hall Company
Reston, Virginia 22090

Library of Congress Cataloging in Publication Data

ANDERSON, LEONARD R (date)
  Electric machines and transformers.

  Includes index.
  1. Electric machinery.   2. Electric transformers.
I. Title.
TK2000.A57   621.31′042   80-18620
ISBN 0-8359-1615-4

CONTENTS

DIRECT-CURRENT GENERATORS 1

DIRECT-CURRENT MOTORS 27

TRANSFORMERS 53

POLYPHASE INDUCTION MOTORS 103

CONSTRUCTION AND THEORY OF OPERATION   104

Construction of a Squirrel-Cage Motor, *104*
Rotating Field in a Two-Pole Motor, *106*

7 SINGLE-PHASE MOTORS 177

THE MECHANICS OF ELECTRIC MOTOR DRIVES 199

GENERAL INFORMATION ON MOTORS AND ELECTRIC POWER SYSTEMS 217

MOTOR STARTERS AND CONTROLLERS 253

APPENDIX:
COST OF ELECTRICAL ENERGY 295

INDEX 297

DIRECT-CURRENT GENERATORS

Since the earliest electrical power systems were direct current (dc), the dc generator was absolutely essential to the operation of those systems. When alternating-current (ac) systems became popular, the use of dc generators declined somewhat, but they were still retained for automotive and aircraft systems and for industrial drives that required the use of a dc motor.

The development of solid-state diodes and silicon-controlled rectifiers (SCRs) has somewhat changed the picture. It is now usually more economical to obtain dc power by rectifying the available ac supply than it is to use an ac–dc motor–generator set. Even on mobile applications, the trend is to generate ac power and rectify to direct current.

There are, however, a fair number of dc generators in service. Technicians and tradesmen must still be prepared to operate, service, and repair these machines, and that is the reason why this chapter has been written.

It should be pointed out that dc generators and dc motors have almost identical construction, and much of the mathematics is also the same. We have therefore tried to keep the construction and mathematics as simple as possible in this chapter. Chapter 2 provides more detail for those who require it.

CONSTRUCTION AND THEORY OF OPERATION

Magnetic Circuit

Figure 1-1 is a sketch of the magnetic circuit for a two-pole dc generator, and the names usually assigned to the various parts are also shown. The field frame or yoke is usually an iron or steel casting of which the inside periphery is machined to ensure accurate positioning of the pole cores. The pole cores are usually built up of silicon steel laminations that are stamped in the appropriate shape, riveted together, and bolted to the yoke. The armature core is basically a cylindrical stack of disc-shaped silicon steel laminations

1

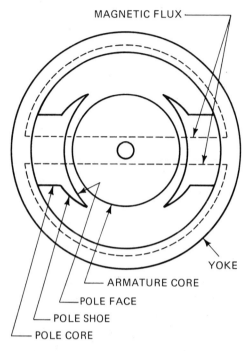

MAGNETIC FLUX

YOKE

ARMATURE CORE

POLE FACE

POLE SHOE

POLE CORE

FIGURE 1-1   Magnetic circuit for a two-pole
generator

mounted on a shaft and supported between the pole cores by means of bearings that will permit it to rotate freely. The space between the pole cores and the armature core is known as the air gap and is usually about 3 to 5 millimeters (mm).

A basic requirement for a generator is a strong stationary magnetic field in the space occupied by the armature. The necessary magnetic field can be readily obtained using a magnetic structure like that in Figure 1-1, by feeding a dc current through coils wound around the pole cores. The line of force will follow a figure eight pattern, as shown in Figure 1-1. It is also possible to use permanent magnets for the pole cores, but we will consider these details later. The important thing is the presence of a stationary field flux through the space where the armature is located, and the reader should note that rota-

tion of the armature core does not significantly change the field flux.

Armature Core and Coils

At first glance, an armature core appears to be a solid iron mass. However, it is always laminated, and the laminations are punched and assembled so that the core has slots near the outer periphery, roughly parallel to the shaft. The slots are lined with insulating paper, and the coils are placed into the slots (frequently wound directly therein). On a two-pole machine the two sides of any one coil will be in diametrically opposite (or very nearly so) slots. Let us consider an armature with 27 slots. If we number the slots consecutively around the armature core, an elementary winding will have a coil in slots 1 and 14, 2 and 15, 3 and 16, and so on, around to slots 27 and 13, and all these coils will have the same size of wire and the same number of turns. These coils would be more correctly called elements or single-element coils. Multi-element coils are more common, of course, but we will discuss that detail in Chapter 2 and make the distinction between "element" and "coil" at that time.

Whenever a conductor moves across a magnetic field, a voltage will be generated in that conductor at right angles to the direction of the field and at right angles to the direction of motion. The magnitude of the voltage is always equal to the number of webers of flux cut per second, and the direction of the voltage can be found by using the right-hand rule.[1] In a generator, rotation of the armature (moving the coil sides across the field

[1]Extend the thumb, index finger, and remaining fingers of the right hand at right angles to each other in three planes. If the hand is positioned so that the index finger points in the direction of the flux and the thumb indicates the direction in which the conductor is moving, the remaining fingers indicate the direction of the induced voltage in that conductor.

flux) generates a voltage in both sides of the coil, and these two voltages add together. In a two-pole machine it is apparent that the direction of the voltage in one side of any one coil will reverse direction every half-revolution. If the two sides of any one coil are 180° apart, the voltages in the two sides will add together at all times. However, the voltage in each coil will be alternating voltage, reversing its direction every half-revolution; furthermore, the coil voltages do not all reverse at the same time. To get a useful output, we must find a way to make these coil voltages add together and get them converted to direct current.

Commutator

A commutator is basically an assembly of rectangular (side view), wedge-shaped (end view) copper bars with deep V-shaped notches in each end, as shown in Figure 1-2. These bars are then stacked into a circular formation with pieces of mica insulation between them, and held in place through axial pressure on two insulated steel end rings that have a truncated-V cross section. The commutator is pressed onto the armature shaft, and the

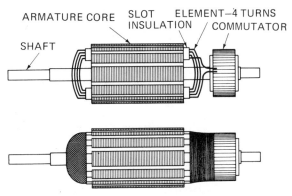

FIGURE 1-3   Two dc generator armatures

outer periphery is then machined to provide a smooth surface to which a stationary carbon brush can maintain continuous contact as the armature and commutator rotate.

The armature elements are connected to the commutator bars, and the brush and commutator assembly does two things:

■ It provides the electrical connection between the rotating armature coils and the stationary external circuit.

■ As the armature rotates, it performs a switching action, reversing the electrical connections between the outside lines and each armature coil in turn so that the armature coil voltages add together and result in a dc output voltage.

Figure 1-3 shows two small dc generator armatures, one with only one element in position, while the other is complete.

Two-Pole Simplex Lap Winding

Figure 1-4 is a sketch of a two-pole simplex lap winding. There are eight slots in the armature core, eight coils or elements, and eight commutator bars. Theoretically, the elements can have any number of turns, but it is convenient to visualize them as having only one.

FIGURE 1-2   Two typical commutator bars

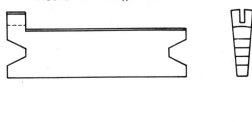

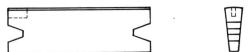

If the connections are carefully traced out, it will be found that from commutator bar 1, the circuit goes through element 1 to bar 2. From bar 2, the circuit goes through element 2 to bar 3, and so on, until finally the circuit goes from bar 8 through element 8 and back to bar 1.

The winding in Figure 1-4 is described as two pole because it is designed for use with a two-pole field structure and would not operate with any other. Lap windings are those in which the two ends of any one armature element are connected to commutator bars that are quite close together, so the winding in Figure 1-4 is described as a lap winding. Because the two ends of each element are connected to adjacent commutator bars, this winding has the minimum possible number of parallel paths and is therefore known as a simplex winding.

Spatial Distribution of Voltage and Current

If the armature in Figure 1-4 is rotated clockwise and provided with a field flux traveling from left to right, the electromotive forces (emf's) induced in the coil sides will be as shown by the crosses and dots[2] within the conductor cross sections. If we trace the circuit from bar 4 to bar 8 through either path, we will find that the emf's produced in the coils that are in series in those paths add together. It therefore follows that the highest voltage will be observed between bars 4 and 8. If we wait until the armature has rotated 45° clockwise from its present position, it is

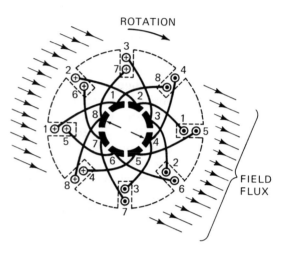

FIGURE 1-4   Two-pole simplex lap winding

apparent that bar 3 will be where 4 is now and bar 7 will move in where 8 is located; but the highest voltage will now appear between bars 7 and 3. To get the highest possible output voltage, we must therefore position the brushes where the two arrows (shown inside the commutator) are located and let the armature spin. If the brushes touch the commutator at the arrows, then, as seen from the external circuit, the coil voltages add together, and the connections between each coil and the external circuit reverse at the same time (or armature position) that the coil voltage reverses.

The spatial positions of the voltages inside the armature are determined by the positions of the field poles, and that relationship cannot be changed. We could shift the connections between the armature elements and the commutator bars, but that would necessitate a new brush position.

If we had a current flow in on the left-hand arrow and out of the armature on the right-hand arrow, the crosses and dots shown would also correctly represent the current flow in the individual coil sides. If the armature in

___

[2]The cross indicates that the emf is away from the observer. The dot indicates that the emf is toward the observer. The same symbol may be used for currents. As a memory-jogging device, think of an arrow. If you see the point (dot), it is coming toward you. If you see the tail feathers (cross), it is going away.

Figure 1-4 rotates so that bars 7 and 3 are in contact with the brushes, a "short-circuit-and-reverse" kind of switching action occurs with regard to coils 7 and 3, and from a stationary reference point of view the spatial distribution of current around the armature core is unchanged. The only way in which the spatial positions of the armature current can be changed is either to move the brushes or shift the connections between the armature elements and the commutator bars (i.e., connect element 1 to bars 2 and 3, element 2 to bars 3 and 4, etc.). We can therefore take a simplified view of an armature.

Figure 1-5 shows two simpler representations of a two-pole armature. The large circle reminds us of the general (end view) shape of an armature, and we show two rectangular brushes in contact with it. The small circles (if they appear) represent the armature conductors, and we may show the direction of current flow or voltage induced in them. However, if the armature conductors are not shown, the current flow is always one way on the left side of the brushes and the other way on the right side. These symbols for an armature may be turned 90° or 180° if convenient.

The position in which the brushes are

FIGURE 1-5 Two simplified representations of an armature

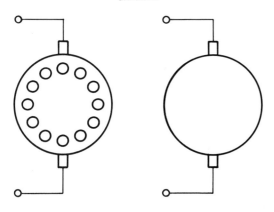

shown needs some explanation. We really show the brushes in the positions where the element sides will be when the short-circuit-and-reverse switching action (i.e., commutation of each new element) occurs. Students sometimes get the idea that the brushes (as a rough approximation) should be about midway between the field poles, and this is almost never the case.

Generated and Terminal Voltage

While the rotation of the armature (which causes the coil sides to move across the magnetic field) induces a voltage in the armature windings, the net voltage obtained from the armature depends on many things, such as the number of turns in the coils, the number of coils in series between brushes, the speed of rotation, and the strength of the magnetic field. However, once an armature has been wound, only the last two factors are variable, and so we can write Equation 1-1.

$$E_G = K_e\phi \cdot \text{revolutions per second (rev/s)}$$
$$(1-1)$$

where E_G = generated voltage
ϕ = total field flux, in webers
K_e = all other factors (that are constant for any given machine)

Under no-load conditions, the voltage E_G will be observed at the brushes, but when the generator is supplying current to the load, part of this voltage is used to push current through the resistance of the armature windings. As a result, the useful output voltage will be a little smaller than E_G, as indicated by Equation 1-2.

$$V = E_G - I_A R_{AE} \qquad (1-2)$$

where I_A = total armature current, in amperes
V = output or terminal voltage

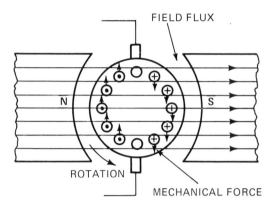

FIGURE 1-6   Torque development in a dc generator

E_G = generated voltage
R_{AE} = effective armature circuit resis-
     tance, in ohms

Torque in a DC Generator

There are two ways to explain how torque is
developed in a dc generator, and both expla-
nations are given here.

If a current-carrying conductor is situated
in an external magnetic field, a mechanical
force will be developed on that conductor at
right angles to its own length and at right
angles to the magnetic field. It is not easy to
explain why that force is produced, and we
will not attempt an explanation, but its presence can be easily demonstrated with a permanent magnet and a coil of wire in free air. The
force on the conductor is always proportional
to the flux density surrounding it and directly
proportional to the current flow in the conductor, and its direction can be found using
the left-hand rule.[3] In a generator the element-

to-commutator connections and the brush
position are always such that all the armature conductors in front of the south pole face
carry current one way, and all the conductors
in front of the north field pole carry current
the other way. The left-hand rule can then be
used to determine the direction of the forces
and net torque produced on the armature, as
shown in Figure 1-6. The conductors in the
spaces between the poles contribute virtually
nothing to the torque.

If desired, one can also relate torque to
the cross-magnetizing effect of the armature.
As shown in Figure 1-7, the flow of armature
current tends to magnetize the armature core,
creating north and south poles on it. If the
brushes are positioned so that the armature
magnetomotive force (mmf) is at right angles
to the field poles, the armature magnetic poles
are attracted to the unlike field poles and
torque is thereby produced on the shaft.

In a generator, the internal emf's are always in the same direction as the current flow,
and if the emf's are as indicated, the rotation
must be clockwise and the torque produced is
opposing the rotation. For this reason the
torque is sometimes known as the countertorque of the generator. The prime mover

[3]Extend the thumb, index finger, and remaining fingers of the left hand at 90° angles to each other in three
planes. If the hand is then positioned so that the index
finger points in the direction of the magnetic field and the
remaining fingers point in the direction of current flow,
the thumb will point in the direction of the force on the
conductor.

FIGURE 1-7   Cross-magnetizing effect of armature
currents

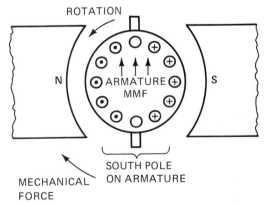

must not only overcome this countertorque; it must also supply enough torque to overcome the rotational losses of the generator, so the actual torque at the shaft (the input or driving torque) must be slightly higher. One important fact is immediately clear. If a generator armature is not supplying any current, it turns easily, but as soon as the generator delivers current, it loads the prime mover.

It is important to remember, too, that the torque is always proportional to the product of the field flux and armature current as stated in Equation 1–3.

$$T = K_T \phi I_A \qquad (1\text{–}3)$$

where T = torque, in newton-meters
   ϕ = flux, in webers
   I_A = armature current, in amperes
   K_T = a constant that takes into account all other factors (such as the number of turns of wire in the armature elements, diameter and length of the core, etc.) that are constant for any given machine

Generator Performance and the Elementary Equations

Equations 1–1 through 1–3 are the three elementary equations that describe generator behavior. We will do a few examples to show how these equations can be used to guide our thinking, but we must remind ourselves of three things. First, the load current is determined by the voltage (V) and the load resistance. One might look at Equation 1–2 and conclude that I_A can be increased by reducing V, but that idea is misleading. The only things that can be conveniently changed on a generator are the speed, the field flux, and the load current. Except for parallel operation, it is best to regard the terminal voltage (V) as dependent on these other factors. The second

point is that V is usually about 90% of E_G. Anything that raises E_G will increase V, also. Third, Newton's law, which states that force equals mass times acceleration, applies to the armature of a dc generator, just as it would to any other object.

What happens if the torque of the prime mover is increased? From Newton's law we know that the generator will accelerate, and from Equation 1–1 we know that, as the speed rises, E_G will rise, and so V increases. If the load resistance is constant, Ohm's law tells us that the load current will increase; therefore, I_A increases and from Equation 1–3 we can see that the countertorque will increase. So the speed will stabilize at a higher value, and the output voltage and current will increase.

How does changing the load resistance affect the generator? If we reduce the load resistance, the load current will increase, so I_A increases. This increases the value of the term $I_A R_{AE}$ in Equation 1–3, so V must go down; but since $I_A R_{AE}$ is small compared to E_G, the reduction of V is moderate and not enough to prevent the increase of load current we originally expected. From Equation 1–3, the torque load on the prime mover is also increased, and usually this results in a modest decrease of speed.

Armature Reaction and Corrective Measures

Until now we have assumed that the field flux goes straight through the armature core as shown in Figure 1-6, and if the armature current is small (e.g., if the generator is running at no load) this assumption is reasonably accurate. However, the flow of current in the armature winding always has a cross-magnetizing effect, as shown in Figure 1-7, and this distorts the field flux as shown in Figure 1-8. This distortion of the main field flux is known as armature reaction. The more armature current we have, the greater the distortion, and in a generator the field flux is always twisted in the direction of rotation.

Armature reaction is objectionable because of the following:

■ It reduces the total field flux. Crowding the field flux toward the trailing edges of the field poles raises the flux density in that vicinity, and because the permeability of iron decreases at higher densities, the constant field mmf produces less flux.

■ It necessitates moving the brushes. Let us define the neutral plane as a line drawn through the center of the armature at right angles to the field flux. To avoid excessive sparking and erosion of the brushes, commutation of each element must occur when the element sides are in the neutral plane, or at least fairly close to it in an area we call the commutating zone. If the field flux distorts, the neutral plane moves, and the brushes must be moved accordingly. Moving the brushes is not very satisfactory, because increased load on the generator causes increased armature current, which

FIGURE 1-8   Field flux distortion due to armature mmf

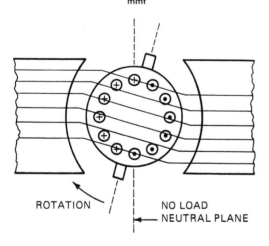

ROTATION   NO LOAD
                NEUTRAL PLANE

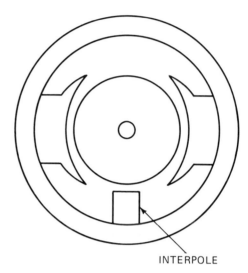

INTERPOLE

FIGURE 1-9   Field structure for a two-pole generator with one interpole

results in more armature reaction. So we have to constantly readjust the brush position to accommodate changes in load. Shifting the brushes also changes the direction of the armature mmf so that it partially opposes the main field mmf and further reduces the total field flux. In fact, if the brushes are moved through a given angle, the armature conductors in twice that angle can be considered as having a demagnetizing effect on the field, reducing E_G.

Motor manufacturers can do four things to minimize armature reaction. Flattening the pole faces slightly so that the air gap is greater at the pole tips than it is at the center of the pole will help. Building the generator with a longer air gap also helps, because this necessitates more field ampere turns to set up the field flux, and the armature ampere turns become smaller by comparison. The machine can also be designed with interpoles and/or compensating windings if necessary.

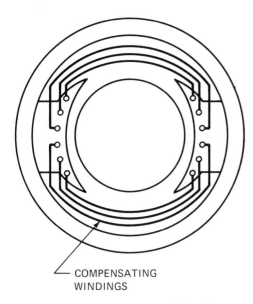

COMPENSATING
WINDINGS

FIGURE 1-10 Compensating windings in a
two-pole field structure

Interpoles or commutating poles are narrow poles located halfway between the main poles on the field structure. They are wound with heavy wire and connected in series with the armature in such a way as to oppose the armature mmf. Interpoles straighten out the field flux in the vicinity of the no-load neutral plane, and therefore eliminate the need for readjustment of brush position to suit load conditions. Field distortion still occurs under the main poles. It is quite common for a two-pole machine to have only one interpole, as shown in Figure 1-9. In a generator, if the interpole is correctly connected, it always has the same magnetic polarity as the preceding main pole (against the direction of rotation).

Compensating windings are rather heavy conductors wound into coils that are placed in slots located in the faces of the main poles, as shown in Figure 1-10. They are connected in series with the armature, and the number of turns is such that their mmf is equal and opposite to the armature mmf so that they will practically eliminate field distortion under the main poles.

It is worth noting that, since interpoles and compensating windings both oppose the armature mmf, once these coils have been correctly connected in series, the connections need not be changed.

Armature reaction is most severe in large and/or high-speed machines. A 0.5-kilowatt (kW), 29 rev/s motor or a 30-kW, 2.2 rev/s design might have neither interpoles nor a compensating winding. But a 2-kW, 29 rev/s motor usually has interpoles, and a 75-kW, 29 rev/s machine will probably have both interpoles and compensating windings.

Commutation and Brush Position

Let us now consider commutation in a little more detail. As each element undergoes commutation, its current must quickly change to an equal value flowing in the opposite direction. Because of the inductance of each coil (and sometimes because of mutual inductance between elements that commutate at the same time), a voltage known as a reactance voltage will be produced in the element during the commutating period. This reactance voltage always opposes the change of element current. Unfortunately, if the current does not reverse properly while the element is shorted, it will have to suddenly jump to the proper value when the associated commutator bars move away from the brush. The high reactance voltage produced by this sudden change of current causes vicious sparking between the trailing edge of the brush and the trailing edge of each commutator bar. Figure 1-12 shows graphically what is required.

There are several armature design features that affect the reactance voltage, and we will consider them on pp. 36–45 of Chapter 2. However, it is important to know what is meant by emf commutation and resistance commutation.

If the brushes are positioned so that commutation occurs slightly behind the neutral plane, or if the interpoles are slightly stronger than the armature mmf, there will be an emf induced in the element while it is short circuited. This emf tends to reverse the current, and if it exactly matches the reactance voltage, sparkless commutation will result. This principle is known as emf commutation.

Resistance commutation is based upon the idea that, as each commutator bar moves out from beneath the brush, the decreasing contact area increases the resistance between that

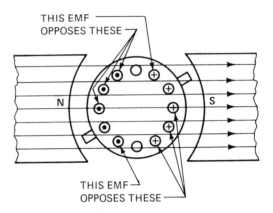

THIS EMF OPPOSES THESE

THIS EMF OPPOSES THESE

FIGURE 1-12   Direct-current generator with the brushes wrongly positioned

FIGURE 1-11   Element current during commutation

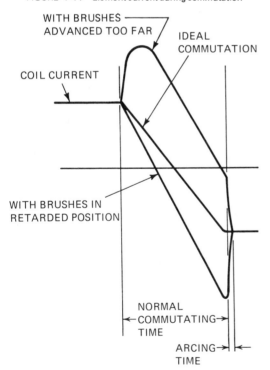

WITH BRUSHES ADVANCED TOO FAR

IDEAL COMMUTATION

COIL CURRENT

WITH BRUSHES IN RETARDED POSITION

NORMAL
←COMMUTATING→
TIME

ARCING→
TIME

bar and the brush. This increase of resistance tends to force the load current to flow through the element that is still (technically) shorted and in this way helps to reverse its current. The higher the contact resistance, the more effective this action will be. High contact resistance also reduces circulating currents (caused by wrong brush position) in the shorted element and thereby makes brush position less critical.

It has already been stated that commutation of an individual armature element must occur when its sides are at or slightly behind the neutral plane. There are two reasons for this. First, if commutation occurs at some other position, some of the armature coils will be connected backward relative to the others. The emf of those coils will oppose that of the others, and the net voltage will be reduced. This is illustrated in Figure 1-11. If the brushes are positioned so that the commutating zone falls at the middle of the field poles, the net voltage becomes zero. Second, if the brushes are wrongly positioned the vicious sparking that occurs at the trailing edges of the brushes and commutator bars is intolerable.

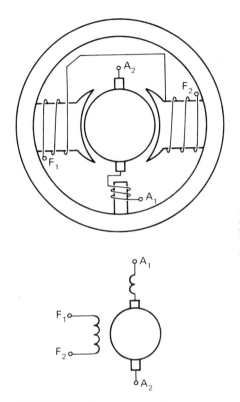

FIGURE 1-13 Separately excited generator

Permanent Magnet Generators

On these machines, the pole cores are permanent magnets, and so the field flux is essentially constant. The only practical way to change the output voltage is to vary the speed.

When load is applied to a permanent magnet generator, the output voltage drops for two reasons. First, current flow through the armature circuit resistance causes V to decrease as shown in Equation 1–2. Second, armature reaction tends to weaken the field flux, causing a decrease of E_G. Large permanent magnet generators are extremely rare. However, under very light load, V becomes very nearly equal to E_G, which is proportional to speed. For this reason small permanent magnet generators (called tachometer generators) are used for speed-measuring purposes.

Types and Characteristics of DC Generators

The most important characteristic of any electrical power source is its voltage regulation. Voltage regulation is defined as the change of voltage that occurs due to a change from full-load to no-load conditions, with all other factors remaining constant. If expressed as a percentage, voltage regulation is given as a percentage of full-load voltage, as indicated by Equation 1–4.

Voltage regulation (%)

$$= \frac{V_{NL} - V_{FL}}{V_{FL}} \times 100 \quad (1\text{–}4)$$

where V_{NL} = no-load voltage
V_{FL} = full-load voltage

We commonly classify dc generators in terms of the method used to obtain the field flux. The various types tend to have distinctive voltage-regulation characteristics.

Separately Excited Generators

Field flux in these generators is set up by current flowing through coils that wrap around the pole cores. These coils are generally wound with comparatively fine wire, but they have a large number of turns so that the necessary field mmf can be created using a fairly small current. The voltage ratings of the armature and field circuits in these machines are not always the same. The internal wiring and the schematic diagram of these generators are shown in Figure 1-13.

The output voltage of a separately excited

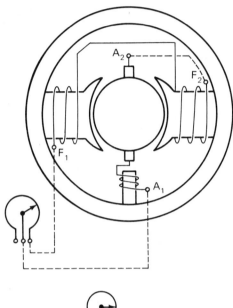

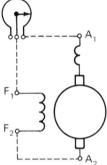

FIGURE 1-14   Shunt-wound dc generator

Self-Excited Generators

The term "self-excited" means that the armature supplies the current to its own field coils. There are four types of these machines.

SHUNT-WOUND GENERATORS   These have a single set of field coils that are connected across the armature, that is, in parallel with the load. The output voltage is usually controlled by connecting a rheostat in series with the shunt field, as shown in Figure 1-14. Increasing the resistance of the rheostat will decrease the output voltage.

The terminal voltage (V) of a shunt generator droops (with increasing load) for three reasons. Armature resistance drop and armature reaction have the same effect here as they have on separately excited machines. However, if the terminal voltage droops, the shunt field current decreases, and this results in less field flux, which causes a further drop in terminal voltage, as shown in Figure 1-15. However, the voltage regulation is not too bad, about 10% to 15%.

We usually think that decreasing load resistance will permit increased load current,

FIGURE 1-15   Drooping voltage characteristics of a shunt generator

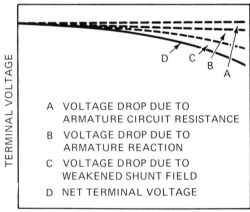

A   VOLTAGE DROP DUE TO ARMATURE CIRCUIT RESISTANCE

B   VOLTAGE DROP DUE TO ARMATURE REACTION

C   VOLTAGE DROP DUE TO WEAKENED SHUNT FIELD

D   NET TERMINAL VOLTAGE

TERMINAL VOLTAGE

LOAD CURRENT

generator can be controlled by varying the speed, but the preferred method is to control the field flux by varying the field current using a rheostat in series with the field. The main advantage of separate excitation is that the output voltage can be adjusted over a wider range than is usually possible with other machines.

When load is applied, armature resistance and armature reaction both cause the output voltage to droop moderately, less than 10%.

and normally this is the case. However, if a shunt generator is sufficiently overloaded, the weakening of the shunt field and the effects of armature reaction become so bad that, if the load resistance is lowered, the generator voltage drops considerably and the load current actually decreases. If we reduce the load resistance to zero (i.e., a dead short circuit right at the brushes), terminal voltage and field current must go to zero. However, residual magnetism will produce a small voltage inside the armature, and so a moderate circulating current will flow under these conditions. Figure 1-16 shows what happens to the the voltage and current as load resistance decreases.

SERIES-WOUND GENERATORS A series-wound generator is shown in Figure 1-17. Under no-load (i.e., open-circuit) conditions the series generator has no field excitation. However, residual magnetism will enable it to generate a small voltage. When the generator supplies load current, this provides field excitation and raises the generated voltage. If the load current goes high enough, saturation of the magnetic circuit will set in,

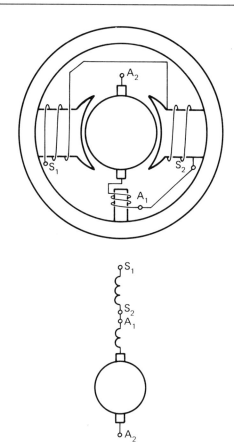

FIGURE 1-17   Series-wound generator

FIGURE 1-16   Voltage–current characteristics of shunt-wound generators

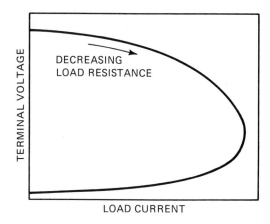

limiting the maximum generated voltage. At the same time, armature reaction and armature resistance drops become significant, and these will cause the terminal voltage to decrease. So the terminal voltage curve (with respect to current) is shaped like that shown in Figure 1-18.

Because its no-load voltage is nearly zero, the series generator is not suitable for general-purpose use. It has been used as an exciter for some welding generators. Its voltage can be reduced by putting a diverter in parallel with the series field.

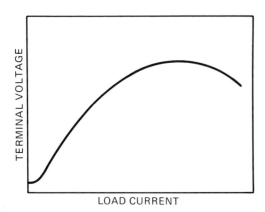

FIGURE 1-18 Terminal voltage characteristics of
a series generator

COMPOUND-WOUND GENERATORS   As
shown in Figure 1-19, these generators have
both shunt and series field coils. Normally,
the shunt field will have the greatest mmf.
Sometimes the field coils are connected so
that the series field carries both the load current and the shunt field current, and this is
known as a long-shunt connection. If the
series field does not carry the shunt field current, the arrangement is known as a short-
shunt connection. Both long- and short-shunt
connections are shown in Figure 1-19. Changing from long to short shunt has little effect
on generator performance.

If the series field is connected so as to aid
the shunt field (a cumulative compound connection), then when load is applied the series
field raises E_G and thereby tends to offset the
internal voltage drops in the machine. The
overall effect is an improvement of voltage
regulation. With a given no-load voltage, the
more series field turns there are the higher
the full-load voltage becomes. If the full-load
voltage is still less than the no-load voltage,
the generator is said to be undercompounded.
If the full-load and no-load voltages are equal,
the machine is described as being flat com-
pounded. If the full-load voltage is higher

than the no-load voltage, the generator is
overcompounded, as shown in Figure 1-20.

An overcompound generator is capable of
maintaining nearly constant voltage at a re-

FIGURE 1-19 Long- and short-shunt compound
generators

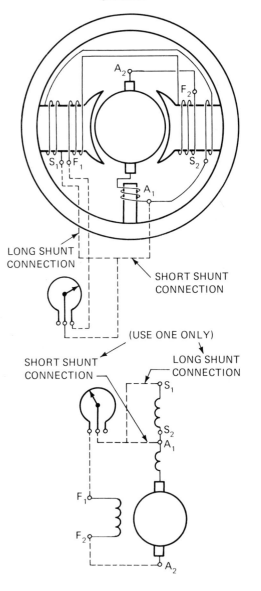

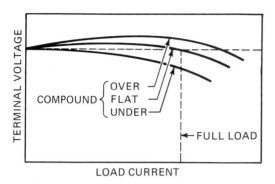

FIGURE 1-20 Terminal voltage characteristics of cumulative compound generators

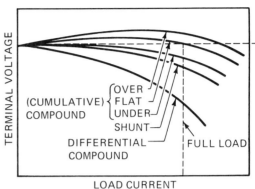

FIGURE 1-21 Terminal voltage curves with equal no-load voltages

mote point on the system it supplies. Some machines have tapped series field coils so that the degree of compounding can be adjusted. However, it is nearly always possible to reduce the degree of compounding by connecting a diverter (a low resistance) in parallel with the series field.

An interesting control problem can arise with cumulative compound generators, particularly if they are overcompounded. Retentivity in the field circuit means that, to bring the generator voltage to a given value, you will require less ampere turns to move down to that value than you will require to bring the voltage up to that value. Because of the series field ampere turns, if the shunt field rheostat does not have enough resistance, it may be found that, once the generator voltage exceeds a certain level, it is impossible to bring the voltage down using only the shunt field rheostat. The immediate solution is to reduce the generator speed, but the permanent cure is to either reduce the degree of compounding (e.g., abandon the series field) or else to use a shunt field rheostat with more ohms.

In a differential compound machine, the series field opposes the shunt field mmf, and this causes the terminal voltage to decrease rapidly as load is applied. The more turns of

wire there are in the series field coils, the more rapidly the terminal voltage drops, and on some machines the full-load voltage may be only one-third of the no-load voltage. Because they operate at rather low field flux densities, they tend to be rather large for their output power (kilowatt) rating. If shorted at their terminals, differential compound generators circulate only a rather small current.

From a theory standpoint, it seems logical to compare the terminal voltage graphs of the various kinds of generators by assuming the same no-load voltage as shown in Figure 1-21. However, the voltage rating of a generator is normally its full-load voltage, and so the comparison should be made as shown in Figure 1-22.

Parallel Operation of DC Generators

Parallel operation means the use of two or more generators to supply a common load. There are three problems associated with parallel operation of shunt-wound generators. First, the installation must be wired so that when the switches are closed the positive

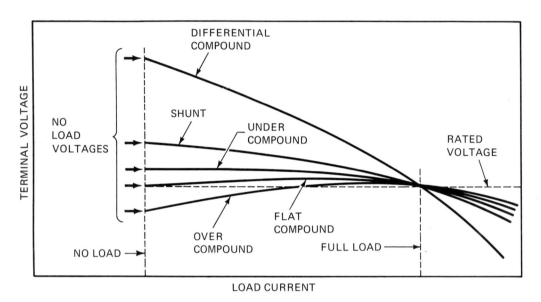

FIGURE 1-22   Terminal voltage curves with equal full-load voltages

terminals of all the generators will be connected to one line and all negative terminals to the other line, as shown in Figure 1-23. The second is an operating problem in that two generators may be tied in parallel only if their voltages are very nearly equal. The third is a design problem in that, while it is possible to adjust the amount of load carried by each generator at any time, we would like to avoid having to make frequent readjustments. Let us examine these problems in more detail.

On the initial installation of a system like that in Figure 1-23, it is fairly easy to check the generator polarity. Start all the generators, close each generator switch one at a time, and check the line polarity with a voltmeter. Reversed generator polarity will be immediately obvious.

When generators are operating in parallel, the power delivered by each can be adjusted by changing its E_G. This could be done by adjusting the speed, but usually we adjust the field excitation. If the total kilowatt load is constant, raising the output of one generator

reduces the load on the others, causing an increase of system voltage. To prevent this voltage change, reduce the excitation (or speed) of the other machines. Disconnecting one generator from the others by arbitrarily opening its line switch would force the other units to suddenly take over its share of the load, causing a sudden decrease of voltage. To prevent this, we first adjust the output of the machine to zero and then open its line switch.

FIGURE 1-23   Three shunt-wound generators connected for parallel operation

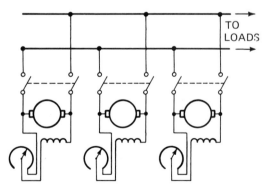

Similarly, to put a generator back on line, we adjust its voltage to match the line voltage and then close the switch. The generator will then be "floating on the line" (i.e., connected but not delivering current), and we can raise its excitation to make it supply power to the system as desired.

For parallel operation we generally want the machines to share the load in direct proportion to their kilowatt ratings. By definition, the terminal voltages have to be equal, and so we adjust the generated voltage of each unit to make it deliver its share of the load. However, if the generators all have the same amount of voltage regulation, this necessitates setting their generated voltages to the same value, and they will then share the load in the same proportion even if the total load changes; this is the preferred arrangement. If the generators do not have the same voltage regulation, the machine with the greatest voltage regulation will have the smallest change of load; so if the total load increases, that machine will not carry its full share unless we readjust its field current. This load-sharing proposition is shown graphically in Figure 1-24. At the initial point of operation, the total load is 200 amperes (A), half of it from each generator. If the total load decreases to 100 A, the terminal voltage rises from 250 volts (V) to 253 V, and the generator currents go to 30 and 70 A, respectively (new points of operation).

There is little that can be done in the field to change the voltage regulation of a shunt generator. Operating at a lower speed and using more field excitation will flatten the voltage curve, but this is not always practical. However, the speed-regulation characteristics of the prime movers will also substantially affect load division, and if they have governors with droop adjustments, the desired load-sharing characteristics can usually be obtained.

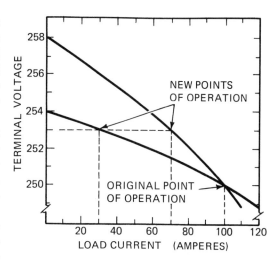

FIGURE 1-24 Load-sharing characteristics of dc generators

Except for the fact that we can use series field diverters to adjust their voltage regulation, compound generators in parallel have all the problems associated with shunt machines; in addition, cumulative types may require an equalizer connection to ensure stable sharing of the load. If the terminal voltage graph of a generator has positive slope with respect to current, any increase of load causes an increase of E_G, so the generator will supply even more current to the system. With two compound generators in parallel, we can end up with one generator supplying all the load current and driving the other generator as a differential compound motor. The equalizer connection overcomes this problem.

The equalizer connection puts the series fields in parallel with each other, as shown in Figure 1-25. A momentary change of load distribution between the generators does not affect the series fields, and so stable operation is obtained. However, the operator must be careful not to open the equalizer connection on a generator unless at least one of the line connections to that machine has already been opened.

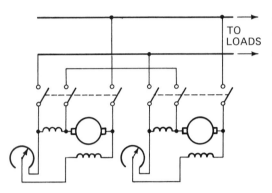

FIGURE 1-25 Parallel operation of compound
generators

Voltage Buildup in Self-Excited Generators

When we rotate a self-excited dc generator, if there is no field flux, the generated voltage will be zero; therefore, it cannot supply any current to its field coils in order to provide a field flux. So while its operation can be readily understood, the question arises, "How does the generator start to generate?" We will answer that question in terms of a shunt generator, but essentially our answer will apply to series and compound types as well.

For a shunt generator to build up voltage, there must be some residual magnetism in its field poles, and the field must be connected to the armature in such a way that the flow of field current will aid the residual magnetism rather than oppose it. When the generator rotates, the residual magnetism enables it to generate a small voltage, which sends current through the field, providing more flux, which raises E_G and field current, and so on, and the whole process is cumulative until magnetic saturation causes the voltage to level off. But let us examine this process in more detail.

Let us take a shunt-wound generator and measure and record the terminal voltage for various values of field current. If we graph

the results of our measurements, we will obtain a graph known as the no-load saturation curve of the generator, and it is always shaped something like that shown in Figure 1-26. If the field circuit resistance is known (shunt field coil resistance plus rheostat resistance), we can also determine the voltage required for various values of field current and plot the results on the same piece of graph paper (this always comes out as a straight line through the origin, and the lower the field circuit resistance, the less slope it will have). Let us call this straight line the field circuit voltage requirement. The generator voltage can stabilize at any point where the no-load saturation curve intersects the field circuit voltage requirement, with the slope of the saturation curve being the lesser of the two. If at a given value of field current the saturation curve falls below the field circuit voltage requirement, the generator voltage must decrease. If the saturation curve is higher than the field circuit voltage requirement, the generator voltage will increase.

Normally, a shunt generator runs quite high on the saturation curve, and it is fairly easy to see how the field rheostat can change the field circuit voltage requirement and thereby give us a new operating point. But there are two other things we can see from Figure 1-26. First, if we add enough resistance to the field circuit, we can bring the field circuit voltage graph back (i.e., swing it counterclockwise) until it is close to and almost parallel with the saturation curve. This makes the generator voltage rather unstable, a commonly encountered problem when attempting to run a shunt generator at well below its rated voltage. Second, if the field circuit resistance is high enough, we get a point of intersection that is quite close to zero. To the casual observer, it appears that the generator is not working at all. Too much field circuit resis-

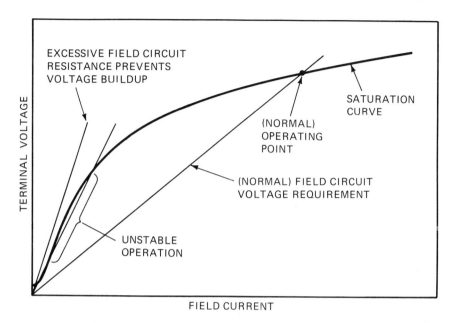

FIGURE 1-26   No-load saturation curve and field circuit voltage requirement

tance is often the reason why generators refuse to build up. Anything that lowers the saturation curve (such as a heavy load on the generator or operation below rated speed) can cause the same problem.

If a generator has lost its residual magnetism, we can give it some by momentarily connecting the shunt field to a dc power supply such as a battery. This procedure is known as "flashing the field." The generator will work no matter which way the field is flashed;

but if we reverse the residual magnetism, we will reverse the output line polarity, and on some installations this cannot be tolerated.

If we reverse the rotation of a generator, the voltage generated due to residual magnetism will be reversed. The brush polarity will therefore reverse, which reverses the field current and the generator will not build up. To operate a generator with reversed rotation, reverse either the armature circuit or both the shunt and series fields.

MATHEMATICAL INTRODUCTION TO DC GENERATORS

Generated Voltage in a DC Machine

It is easiest to develop a generated voltage or emf equation by assuming that the flux per

pole is known and first considering the (absolute) average voltage generated in one conductor. If ϕ represents the flux per pole in webers and P represents the number of poles, then $P\phi$ represents the flux (webers) cut by any one armature conductor in one revolution. Since the voltage generated per conductor is always equal to webers per second, we get Equation 1–5.

$$E_{\text{AVG for 1 cond}} = P\phi \times (\text{rev}/\text{s}) \quad (1\text{–}5)$$

To get the voltage generated in any one path between brushes, we simply take Equation 1–5, multiply by the number of armature conductors, and divide by the number of parallel paths, as shown in Equation 1–6.

$$E_G = \frac{P\phi N_A}{b} (\text{rev}/\text{s}) \qquad (1\text{–}6)$$

If the only information we have is the average flux density under the poles, the length of the armature, and the percentage of the armature periphery covered by the poles' faces, we can obtain the value of $P\phi$ using Equation 1–7.

$$P\phi = \pi dlB \times (\% \text{ coverage}/100) \quad (1\text{–}7)$$

where d = armature diameter, in meters
  l = length of armature core, in meters
  B = flux density, in teslas or webers per square meter

The proper value for N_A also needs explanation. If we could slice an armature at right angles to the shaft and count the number of wire ends visible in the slots, that count would give us the total number of armature conductors, which we will denote as N_A. A little thought will show the validity of Equation 1–8.

$$N_A = 2 \times (\text{no. of elements}) \atop \times (\text{turns}/\text{element}) \qquad (1\text{–}8)$$

EXAMPLE 1–1

An armature that has 240 three-turn elements (total) in 6 parallel paths has a diameter of 0.35 meters (m) and its length is 0.9 m. If the main field poles cover 65% of the armature periphery, and the average flux density under those poles is 0.93 weber (Wb)/m², find the generated voltage when the speed is 3.6 rev/s.

Solution
From Equation 1–7,

$$P\phi = \pi dlB \times (\% \text{ coverage}/100)$$
$$= 3.1416 \times 0.35 \times 0.9 \times 0.93 \times 0.65$$
$$= 0.5982 \text{ Wb}$$

From Equation 1–8,

$$N_A = 2 \times 240 \times 3 = 1440 \text{ conductors}$$

From Equation 1–6,

$$E_G = \frac{P\phi N_A \ (\text{rev}/\text{s})}{b}$$
$$= \frac{0.5982 \times 1440 \times 3.6}{6}$$
$$= 517 \text{ V} \qquad \text{(Answer)}$$

Torque in a DC Machine

The mechanical force produced on a current-carrying conductor that is surrounded by an external magnetic field is given by Equation 1–9.

$$f = Bli \qquad (1\text{–}9)$$

where f = force, in newtons
  B = flux density, in teslas
  i = instantaneous current, in amperes

For a dc generator, we will consider the current in the armature as being constant. We therefore only have to consider how many armature conductors are in front of the field pole faces (these are the only conductors that produce torque) and the diameter of the armature core, and then apply Equation 1–9. The whole operation can be done in one step, as shown in Equation 1–10.

$$T_D = BlIn_A \frac{d}{2} \qquad (1\text{–}10)$$

where T_D = torque developed, in newton meters
  B = average flux density under the pole faces, in teslas

l = length of the armature core, in meters

I = current per path in the armature winding, in amperes

n_A = number of armature conductors that are in front of the pole faces

d = diameter of the armature, in meters

Equation 1–10 is easily remembered as "blind/2" but must be applied with care. Be sure to use the correct value for I as given by Equation 1–11.

$$I = \frac{I_A}{b} \qquad (1-11)$$

where I = current per path

I_A = total armature current

b = number of parallel paths in the armature winding

In a practical dc generator, the main field poles do not cover the whole armature periphery but only about 70% or so. The correct value for n_A in Equation 1–10 is therefore given by Equation 1–12.

$$n_A = N_A \times \frac{\% \text{ coverage}}{100} \qquad (1-12)$$

EXAMPLE 1-2

If the total current through the armature in Example 1-1 is 90 A, find the torque developed.

Solution

From Equation 1–11,

$$I = \frac{I_A}{b} = \frac{90}{6} = 15 \text{ A}$$

From Equation 1–12,

$$n_A = 1440 \times \frac{65}{100}$$

$$= 936$$

From Equation 1–10,

$$T = 0.93 \times 0.9 \times 15 \times 936 \times \frac{0.35}{2}$$

$$= 2056.5 \text{ newton-meters (Answer)}$$

Generated Voltage, Terminal Voltage, and Voltage Regulation

Voltage regulation has already been defined, and the relationship between E_G and V has also been given, so let us do a sample calculation.

EXAMPLE 1-3

A shunt generator that delivers 30 A to the load and 4 A to its own field has a full-load terminal voltage of 250 V. Its effective armature resistance is 0.4 ohms (Ω). If armature reaction depresses E_G by 12 V, and the weakened shunt field causes a further 10-V drop, find the no-load voltage and the voltage regulation (neglect the change of field current).

Solution

$$\begin{aligned}
\text{Full-load terminal voltage} &= 250 \text{ V} \\
\text{Armature reaction} &= 12 \text{ V} \\
\text{Weakened field} &= 10 \text{ V} \\
\text{Change of } I_A R_{AE} = (34 - 4)(0.4) &= 12 \text{ V} \\
\text{No-load terminal voltage} &= \overline{284 \text{ V}}
\end{aligned}$$

$$\text{Voltage reg.} = \frac{284 - 250}{250} \times 100 = 13.6\%$$

(Answer)

Input, Output, Losses and Efficiency

The mathematical relationships between input, output, losses, and efficiency are given by Equations 1–13 and 1–14, which are valid for any machine.

$$\text{Input} = \text{output} + \text{losses} \qquad (1-13)$$

$$\eta = \frac{\text{output}}{\text{input}} \qquad (1-14)$$

where η = efficiency.

The losses of a dc generator are as follows:

Rotational Losses

MECHANICAL LOSSES   The mechanical losses are just windage and friction in the bearings and at the brushes. We will consider these losses to be constant.

IRON LOSSES   Iron losses are composed mainly of hysteresis and eddy current losses in the armature core. Raising either the armature speed or the field flux density will increase these losses. However, given the way that most generators are used, iron losses tend to be constant, and in this chapter we will consider them to be so.

Copper Losses

SHUNT FIELD LOSS   This is equal to the product of the shunt field current and the voltage at the shunt field terminals. Do not use line voltage for this calculation unless it is desired to include the field rheostat losses. If the field rheostat is not moved, these losses remain constant.

ARMATURE CIRCUIT COPPER LOSSES   Included here is the I^2R loss in the armature winding, the brushes, and the interpoles and compensating windings if they exist. If it is a long-shunt generator, the series field would also be included. In this chapter we will lump all these losses together and calculate them using $I_A^2 R_{AE}$. Here is a sample calculation:

EXAMPLE 1-4

A shunt generator rated at 40 A, 125 V has 500 W of mechanical losses, 300 W of iron loss, and requires 3.6-A field current. Its armature circuit resistance is 0.19 Ω. Assuming no change of speed or field current, find the following:

a. The torque required to turn the generator at no load and normal excitation.
b. The power required to turn the generator at no load if we change to separate excitation.
c. The driving power required with no excitation.
d. Its full-load efficiency as a shunt generator (including the field rheostat losses).

Solution

a. At no load, driving power equals losses. In this case, we have

$$
\begin{array}{lr}
\text{Mechanical losses} = & 500\ \text{W} \\
\text{Iron losses} = & 300\ \text{W} \\
\text{Shunt field circuit loss,} & \\
\quad 3.6 \times 125 = & 450\ \text{W} \\
\text{Armature circuit loss,} & \\
\quad 3.6^2 \times 0.19 = & \underline{2.46\ \text{W}} \\
\text{Total} = & 1252.46\ \text{W}
\end{array}
$$

(Answer)

b. With separate excitation, the armature circuit losses go to zero and the shunt field losses are not supplied via the shaft. The driving power required is therefore equal to the rotational losses, which is 800 W.

(Answer)

c. With no excitation all we have left is the mechanical losses, which totals 500 W.

(Answer)

d. At full load,

$$
\begin{array}{lr}
\text{Output} = 40 \times 125 = & 5000\ \text{W} \\
\text{Rotational losses} = & 800\ \text{W} \\
\text{Shunt field circuit loss,} & \\
\quad 3.6 \times 125 = & 450\ \text{W} \\
\text{Armature circuit losses,} & \\
\quad 43.6^2 \times 0.19 = & \underline{361.2\ \text{W}} \\
\text{Input} = & 6611.2\ \text{W}
\end{array}
$$

$$
\therefore \text{Efficiency} = \frac{5000}{6611.2} = 0.7563
$$

(Answer)

SUMMARY

The basic theory of a dc generator is not particularly difficult. If we steer clear of the details of the armature winding, its operation can be readily visualized in terms of simple physical principles and concisely expressed in terms of three elementary equations.

The most important single characteristic of a dc generator is its voltage regulation. To a major degree, the degree of voltage regulation depends upon the method used to obtain the field flux.

Direct-current generators have been manufactured for many years, and their design has been refined to a very high degree. Many aspects of their design (both electrical and mechanical) have not been included here, and readers who wish to pursue these aspects must look to other sources of information.

QUESTIONS

1-1. Explain what is meant by the following terms:
  (a) Armature core
  (b) Commutator
  (c) Field pole
  (d) Yoke
  (e) Pole face
  (f) Simplex
  (g) Generated voltage
  (h) Terminal voltage
  (i) Voltage regulation
  (j) Tachometer generator
  (k) Diverter
  (l) Shunt wound
  (m) Series wound
  (n) Compound wound
  (o) Differential compound
  (p) Unstable
  (q) Parallel operation
  (r) Equalizer connection
  (s) Voltage buildup
  (t) Interpoles
  (u) Compensating windings
  (v) Losses
  (w) Efficiency
  (x) Reactance voltage

1-2. When load is applied, a generator becomes hard to turn. Explain why.

1-3. Explain what will happen if we increase the field flux in a generator.

1-4. What is the most common use of a permanent magnet generator?

1-5. What is the main advantage of a separately excited generator?

1-6. Draw a diagram of a generator in the following three steps:
  (a) Draw a two-pole simplex lap-wound armature that has six slots in the core, six coils, and six commutator bars (similar to Figure 1-4).
  (b) Draw the field poles and arbitrarily assign them their magnetic polarity.
  (c) Assuming counterclockwise rotation, show the direction of the emfs in the coil sides, determine the correct brush position, and indicate which brush is positive.

1-7. On the initial installation of a cumulative compound generator, care must be taken to connect the series field correctly. Why?

1-8. What is the purpose of an interpole?

1-9. Why is it unnecessary to provide separate external leads for the armature, interpoles, and compensating windings on a generator?

1-10. On some generators the brushes must be set so that commutation occurs slightly behind the neutral plane. Why?

1-11. How does brush contact resistance help to reverse the current in an element that is undergoing commutation?

1-12. What three factors cause the voltage regulation of a shunt generator?

1-13. If shorted at its terminals and then

started, a shunt generator delivers very little current. Why?

1–14. Why is the series generator unsuitable for general-purpose use?

1–15. Why does cumulative compounding improve voltage regulation?

1–16. Distinguish between flat, over-, and undercompound generators.

1–17. To obtain proper control, cumulative compound machines require more resistance in the field rheostat than shunt machines. Why?

1–18. Why are differential compound machines usually larger than cumulative types with the same kilowatt rating?

1–19. Why is Figure 1-22 considered preferable to Figure 1-21?

1–20. When generators are operated in parallel, there is a specific procedure to be followed when putting a generator on line. What is the procedure and why is it necessary?

1–21. If generators are operated in parallel, it is preferable that they all have the same voltage regulation. Why?

1–22. Why is an equalizer connection needed for cumulative compound generators in parallel and how does it ensure stable operation?

1–23. For compound generators in parallel, the series fields must all be connected to the positive (or else all to the negative) line. Why?

1–24. Which of the following conditions will prevent a long-shunt compound generator from building up its voltage?

(a) Open shunt field
(b) Shorted series field
(c) Reversed series field
(d) Speed too high
(e) All brushes lifted off the commutator
(f) Reversed rotation
(g) Reversed shunt field
(h) Shorted rheostat

1–25. On a generator that operates in parallel with others, one must take particular care if it becomes necessary to flash the field. Why?

MATHEMATICAL PROBLEMS

1–1. If a generator is producing 125 V, what will its voltage be if
(a) The flux is increased 20%.
(b) The speed is decreased 25%.
(c) Both changes (a) and (b) occur.

1–2. If the countertorque of a generator is 500 newton-meters, find the countertorque if
(a) The current is tripled.
(b) The field flux is cut in half.
(c) Both changes (a) and (b) occur.

1–3. A generator with 1.6-Ω armature circuit resistance has a terminal voltage of 125 V when the armature current is 8 A. Assuming that the generated voltage does not change, find the terminal voltage when the armature current is 18 A.

1–4. Find the armature current for each of the following generators:
(a) Permanent magnet generator, load current of 20 A.
(b) Shunt generator, shunt field current of 20 A, load current of 400 A.
(c) Long-shunt compound generator, series field current of 52 A, load current of 46 A.
(d) Separately excited generator, shunt field current of 3 A, load current of 23 A.

(e) Short-shunt differential compound generator, series field current of 38 A, shunt field current of 3.8 A.

1-5. Find the voltage regulation (%) for each of the following cases:

(a) Full-load voltage of 125 V, no-load voltage of 135 V.

(b) No-load voltage of 490 V, full-load voltage of 430 V.

(c) No-load voltage of 280 V, voltage decrease with load of 30 V.

(d) No-load voltage of 140 V, voltage rise when load is removed of 20 V.

(e) Full-load voltage of 130 V, no-load voltage of 120 V.

1-6. Find the input, output, losses, and/or efficiency as appropriate for each of the following cases:

(a) 600-kW input, 560-kW output.

(b) 75-kW input, 4.2-kW losses.

(c) 15-kW input, 87% efficiency.

(d) 300-W output, 71% efficiency.

(e) 3-kW output, 700-W losses.

(f) 900-W losses, 83% efficiency.

1-7. The no-load voltage of a certain shunt generator is 130 V. Armature reaction at full load causes a decrease of 6 V, and weakening of the shunt field causes an additional 5-V drop. If the armature circuit resistance is 0.14 Ω and full-load armature current is 30 A, find the full-load voltage.

1-8. Find the full-load efficiency of the following long-shunt generator:

Rated output = 40 kW at 250 V

Shunt field resistance = 24 Ω

Shunt field rheostat resistance = 8.3 Ω

Armature circuit resistance = 0.10 Ω

Iron losses = 695 W

Mechanical losses = 840 W

DIRECT-CURRENT MOTORS

Electrical distribution systems are primarily designed to transmit energy from place to place, but the ultimate consumer usually would not want that energy if it had to remain in electrical form. Instead, the consumer generally wants heat, light, and/or motive power, and so appropriate energy conversion equipment has been developed.

The need for motive power has prompted the development of various kinds of electric motors. Since the early electric power systems were direct current, the first motors were dc types. With the advent of ac power systems, the popularity of the dc motor declined, mainly because of its higher cost and its need for more frequent and careful maintenance. However, it never entirely disappeared, partly because the dc motor was inherently the most suitable for smooth, efficient, wide-range speed control, and partly because it was the only type that could be used for automotive and aircraft applications.

With the electrical utilities transmitting energy on alternating current, the usual arrangement once was to have an ac motor drive a dc generator, which would in turn supply a dc motor. However, the advent of solid-state diodes and silicon-controlled rectifiers eliminated the need for the motor–generator set and made dc motor drives more versatile and more popular than ever before.

In the next section of this chapter we will develop the theory of dc motor operation, and this material is a necessary prerequisite for the sections that follow, in which we will treat armature winding designs in more detail so that the reader will understand the circuits or winding patterns in common use. Then we will examine some of the mathematical considerations that form the basis of dc motor theory. The whole chapter can serve either as a fairly complete treatment of dc motors for electrical tradesmen or as a basis for further study along the lines of dc motor design. Motor characteristics, application, and control are treated more fully in Chapters 9 and 10.

CONSTRUCTION AND THEORY OF OPERATION

Motors Compared to Generators

The fundamental difference between motors and generators is the purpose that they serve. A generator converts mechanical power into electrical power, whereas in a motor the conversion is the other way. Motors and generators basically use the same field frames, field coils, armature cores, commutators, brushes, and bearings, and the armature windings follow the same patterns. It is possible to build a machine that will operate satisfactorily as either a motor or generator without alteration. It is possible, however, to make some small refinements of design to get improved operation as a generator (at the expense of motor performance), or vice versa.

Equations 1–1 and 1–3 for generators are also true for motors. However, E_G is usually called the counter emf (cemf) because its direction is always such as to oppose the flow of normal armature current. If the line polarity and all other factors remain unchanged, the applied voltage is normally greater than the cemf. The armature and line currents are therefore both reversed compared to a generator, and this reversal has five significant effects.

First, since the shunt field current does not reverse, the line current must equal the sum of the armature and shunt field currents, as shown in Figure 2-1, and the product of line current and voltage is therefore the total power input.

Second, the reversed armature current causes the armature reaction to be reversed, as shown in Figure 2-2. Instead of being pushed toward the trailing edge of the field poles, the flux gets twisted in the direction opposite to the armature rotation, and so on machines without interpoles, the brushes must be shifted the same way. Interpoles and compensating windings are commonly used on motors, and they serve the same purpose as they do on generators. Their magnetic polarity must be reversed compared to motors, but since the armature current is also reversed, the correct polarity is automatically obtained with no change of connections. Third, on generators it is sometimes helpful to position the brushes so that commutation occurs slightly behind the neutral plane. This technique works on motors, too, but because the armature current is backward by comparison, we must move the brushes the other way, that is, slightly ahead of the neutral plane in the direction opposite to the rotation.

Fourth, the armature current equation is

FIGURE 2-1   Current flow in motors and generators

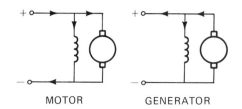

MOTOR   GENERATOR

FIGURE 2-2   Armature reaction in a motor

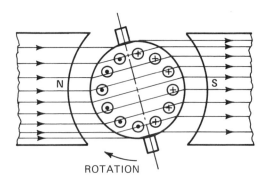

ROTATION

usually written differently for generators and motors. Since E_G for a generator is normally greater than line voltage, we use Equation 2-1 (rather than Equation 1-2).

$$I_A = \frac{V - E_G}{R_{AE}} \qquad (2\text{-}1)$$

Fifth, if a dc machine changes from generator to motor operation, the current in the series field is reversed, and so a cumulative compound generator becomes a differential compound motor, and vice versa.

Motor Performance and the Elementary Equations

The elementary equations for a motor are Equations 1-1, 2-1, and 1-3. They are most useful in determining what will happen if we make some change to an existing motor. We will do several examples to show how helpful these equations are, but to do so we need two additional pieces of information.

First, we must remember Newton's second law of motion, which states that acceleration of an object is directly proportional to the net force on that object. If we apply this idea to a motor and its load, we recognize that to run at a constant speed the motor torque must be equal to the load torque (including the friction in the motor) so that the net torque acting on the assembly will be zero. If the motor torque is greater than the load torque, the machine will accelerate, and if the motor torque is less than the load torque, the machine decelerates.

The second point is that under normal running conditions the cemf of most motors is about 80% to 90% of the applied voltage. What happens if we increase the mechanical load on a motor? From Newton's law we know it will decelerate, but from Equation 1-1 we can see that this will decrease the cemf in the motor. Equation 2-1 tells us that the armature current will increase, and Equation 1-3 shows that this will increase the motor torque. We can therefore see that the speed will stabilize at a new and lower value.

What happens if we raise the voltage applied to the armature and all other factors remain constant? Equation 2-1 shows that we will get more armature current; Equation 1-3 shows that the torque will increase, and therefore the motor will accelerate. However, as the motor gains speed, the cemf will rise, reducing the armature current and torque. The speed will therefore stabilize at a higher value.

The effect of decreasing the field flux is not quite so obvious, because (from Equations 1-1 and 2-1) it will result in a simultaneous increase of armature current, and the overall effect on the torque is therefore uncertain. However, it can be shown that, as long as the cemf does not become less than half of the applied voltage, the percentage increase of armature current will be greater than the decrease of flux, and therefore the torque increases. As a result, the motor accelerates, stabilizing at some higher speed.

Many other things can be learned by examining the basic equations. Here are three examples:

- The armature current on starting (before it has had the time to accelerate) will be much higher than the normal current when the motor is up to speed.
- The no-load current of a motor is always lower than the current with load.
- If we drive a motor above its inherent no-load speed by some external means, the cemf will become greater than the applied voltage, reversing the armature current and torque. It really operates as a generator under these conditions.

Types and Characteristics of DC Motors

Because the torque–speed characteristics are determined largely by the method used to obtain the field flux, dc motors are generally classified according to their field structure and/or circuitry. The characteristics of dc motors are treated in more detail in Chapter 9. Here we will simply list the various kinds of motors, point out their starting torque and speed-regulation[1] characteristics, and briefly explain why each motor behaves the way it does.

Permanent Magnet Motors

The pole cores in these motors are permanent magnets and field coils are therefore not required. However, some of these motors do have coils wound on the poles. If they exist, these coils are intended only for recharging the magnets in the event that their strength is lost.

Because the field flux is constant, the torque of these motors is directly proportional to the armature current, and if the latter is permitted to go high enough, the torque can be very high. Some manufacturers quote 700% starting torque for these designs.

Because the armature resistance is low, a small decrease of speed (and therefore cemf) permits a large increase of armature current (and therefore torque), and so the speed regulation is quite good, about 10% to 15%.

Schematically, this motor appears as only an armature.

Separately Excited Motors

These have field coils similar to those of a shunt-wound motor, but the armature and

field coils are fed from different power sources and may have different voltage ratings. The torque and speed-regulation characteristics are the same as for shunt-wound machines.

Shunt-Wound Motors

The word "shunt" means "parallel." These motors are so named because they basically operate with the field coils connected in parallel with the armature. The field current is much less than the armature current, sometimes as low as 5%.

To facilitate speed control and reversal of rotation, a general-purpose shunt-wound motor normally has four external leads, as shown in Figure 2-3. However, if the motor is built

FIGURE 2-3   Schematic and wiring diagram of a shunt motor

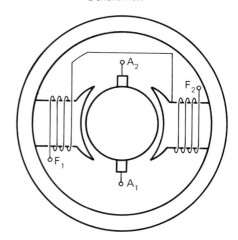

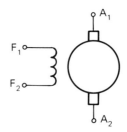

[1]See Chapter 8 for definitions of starting torque and speed regulation.

for some specific application and reversed rotation will not be required, one end of the shunt field may be permanently connected to one armature lead; if neither reversed rotation nor speed control is required, the armature and field may be permanently connected in parallel and only the two external leads provided.

At first glance, one would expect the field flux to be constant in a shunt motor so that its characteristics would be the same as the permanent magnet types. However, the flow of current in the armature slightly reduces the

FIGURE 2-4   Schematic and wiring diagram of a series motor

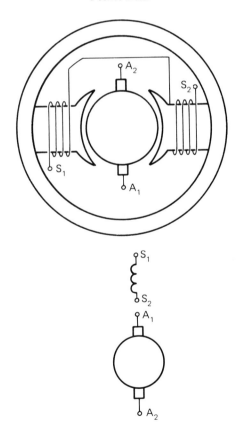

field flux at full load so that the speed regulation is usually less than that of permanent magnet types, sometimes less than 10%. If the armature current is allowed to go to very high values on starting, it tends to reduce the field flux so much that it places a ceiling on the torque that the motor can develop. So a shunt motor is considered to have only a moderate starting torque, 125% to perhaps 200%, using armature current of about the same percentage values.

There is one hazard associated with the operation of shunt (or separately excited) motors. If the field circuit becomes open, the retentivity of the magnetic circuit will provide some residual magnetism, and if the mechanical load is not too high, the motor will accelerate to several multiples of its normal speed. We say the motor runs away, and, while a very small motor might stand this, centrifugal force will usually damage the armature of a large machine when this occurs (e.g., soldered leads pull out of the commutator bars, banding wires break, coil ends bend out until they drag on the field coils, or a commutator bar may fly out). The danger posed by the likelihood of flying parts should not be underestimated.

Series-Wound Motors

As the name implies, the field coils of a series motor are connected in series with the armature, as shown in Figure 2-4. The cross-sectional area of the wire used for the field coils has to be fairly large to carry the armature current, but because of the higher current, the number of turns of wire in them will be much lower than that required for a shunt field.

On starting, the high current through the field coils produces a strong field flux, and this combined with the high current in the

armature winding gives the motor high starting torque, approaching 400% torque with
about 200% starting current.

The speed regulation of a series motor is
normally very poor. The no-load speed is
usually several multiples of the rated full-
load speed, high enough for centrifugal force
to damage the armature on any but the very
smallest machines. The reason for this high
no-load speed is that, as the speed rises, the
field flux is weakened, permitting further increases of speed. To ensure that the motor
will not run away, we prefer to couple the
motor directly to its load rather than use a
belt drive.

Compound Motors

A compound motor has both shunt and
series field coils, as shown in Figure 2-5, with
both sets of field coils being used. The shunt
field is normally the stronger of the two (i.e.,
has more ampere turns).

Manufacturers often build a motor with
both shunt and series fields and tell the customer that it may be used as either a shunt or
a compound machine (such motors usually do
not have enough turns of wire in the series
field to operate as a plain series motor). To
permit this and also to facilitate speed control
and reversal of rotation, the motor will usually have two armature leads, two series field
leads, and two shunt field leads.

Compound motors are usually hooked up
cumulatively; that is, the series field is connected so that it adds to the flux produced
by the shunt field. This gives the motor better
starting torque than a shunt motor but not
as good as in a series type. The speed regulation, however, is not as good as in a shunt
machine, typically from 20% to 50%. Some of
these motors have taps in the series field coils
so that the degree of compounding can be
changed. The more turns of wire used in the

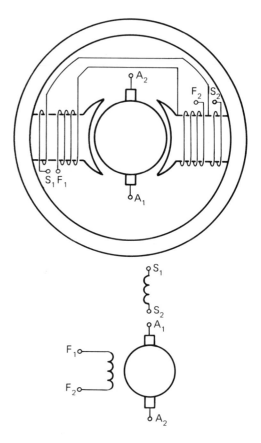

FIGURE 2-5   Schematic and wiring diagram of
a compound motor

series field, the higher the starting torque and
the greater the speed droop as load is applied.
However, the motor always has a definite and
safe no-load speed.

Sometimes we intentionally connect the
series field so that it opposes the flux set up
by the shunt field. This arrangement is known
as a differential compound motor, and its
chief advantage is the excellent (less than 5%)
speed regulation that can be obtained. When
load is applied to the machine, it slows down
a little, reducing the cemf and allowing more
current to flow. However, the increase of cur-

rent in the series field lowers the field flux, and this allows a large increase in armature current and torque with very little decrease of speed.

Differential compound motors have two disadvantages. First, if started as differential machines, the starting torque will be very poor at best. It is therefore almost always necessary to short out the series field during starting. Second, if the series field is too powerful (e.g., too many turns of wire), the motor speed becomes unstable as load is applied. Depending on the relative strength of the shunt and series fields, the motor may run away when it is loaded, or it may stall out; if the series field is strong enough to overcome the shunt field (i.e., reverse the field flux), the motor may suddenly come to a stop and start up with reversed rotation. The current flow and the torque developed during this unstable operation are very high, so it is desirable to avoid this condition. On the initial start-up of a compound motor that is (or may be) differentially connected, short out the series field during starting and, when the motor speed has stabilized, remove the short circuit. If the motor slows down when the short is removed, it is a cumulative connection. If a motor speeds up, it is differential compound. If the motor goes unstable, quickly open the line switch or else replace the short circuit.

It is possible to hook up a compound motor with the series field and armature in series with each other and in parallel with the shunt field. This is known as a long-shunt connection and sometimes results in simpler wiring. However, it is possible to connect the armature and shunt field in parallel with each other and the series field in series with that pair. This is known as a short-shunt connection. Changing from long to short shunt, or vice versa, has little effect on motor performance.

Rotation Reversal and Speed Control

If we carefully consider the right-hand rule given on pp. 6–7, it is apparent that the only ways to reverse the rotation are to reverse the armature current or else reverse the magnetic polarity of the field poles. However, on a motor that has shunt fields, series fields, armature windings, interpoles, and compensating windings, it is difficult to know what to reverse, and so we will go through this in more detail and simplify the problem.

First, the interpoles and compensating windings must carry the armature current only, and they must oppose the armature mmf. It therefore follows that these three components must be directly connected in series with each other; once these connections have been correctly made, they must not be changed, no matter how the motor is to be used. For this reason, most manufacturers make the necessary connections between the armature, interpoles, and compensating winding, and provide only two external leads marked A_1 and A_2. With regard to the shunt and series fields, reversing only one of them would change the motor from cumulative to differential compound (or vice versa). So if a motor is operating properly and it is desired to reverse the rotation, reverse the armature, interpoles, and compensating windings together, or else reverse both the shunt and series fields.

If we examine the basic motor equations with a view toward speed control, we find there are three basic ways to obtain it. Assuming a constant load, the only variables are the field flux, the voltage applied to the armature, and the armature circuit resistance.

Field flux control is practical on all but permanent magnet motors. The stronger the field poles, the slower the motor runs, but saturation of the iron and/or field coil over-heating (due to higher currents) sets a lower limit on the speed that can be obtained. We therefore tend to think of this method as raising the motor speed. On shunt and compound types a rheostat in series with the shunt field is all that is required. On separately excited motors we may use either a rheostat or a variable-voltage power supply. For series motors we may install a low-resistance diverter in parallel with the field (bypassing some of the current to reduce the flux), or the field coils may be tapped so that we can reduce the number of turns. This method does not greatly affect starting torque, speed regulation, or efficiency (in percent).

Armature voltage control is quite practical on permanent magnet, separately excited, and series-wound machines. The basic requirement is a variable-voltage dc power supply, which can be readily obtained by using a dc generator or a silicon-controlled rectifier. The motor speed tends to be directly proportional to the applied voltage, but since it is not practical to raise the voltage much above the armature rating, we think of this method as reducing the motor speed. To provide independent speed control, a separate power supply is required for each motor, but starting torque, speed regulation, and efficiency are not greatly affected. Armature voltage control should not be attempted with shunt or compound machines. Although reducing the armature voltage tends to reduce the speed, a lower applied voltage also weakens the field flux, which tends to raise the speed. As a result, the speed range that can be obtained is rather narrow, and the starting torque is very poor at low speed settings.

Armature resistance control can be applied to any dc motor. All that is needed is to add resistance to the armature circuit and this will necessarily reduce the motor speed. However, armature resistance control has little effect on the no-load speed, so the speed regulation becomes very poor. Starting torque is also lowered, and efficiency is reduced almost in direct proportion to the speed.

See Chapter 9 for more information on starting and speed control of dc motors.

Speed, Torque, and Power Relationships

If we consider a force (f) measured in newtons (N) acting at right angles to a radius of r meters (m), the work done (joules, J) per revolution is given by Equation 2–2.

$$\text{Work} = \text{force} \times \text{distance}$$
$$= f \times 2\pi r$$
$$= 2\pi fr \qquad (2\text{–}2)$$

Since force times radius equals torque, we can make that substitution, and if we multiply Equation 2–2 by revolutions per second, we end up with joules per second, which is just watts (W). We can therefore write Equation 2–3.

$$P = 2\pi T(\text{rev/s}) \qquad (2\text{–}3)$$

where P = power, in watts
T = torque, in newton-meters

For Equation 2–3, one could use developed torque and find developed power (or vice versa) or use output torque and find output power (or vice versa). The difference is just the rotational losses of the motor (which can be expressed either as a torque load or in terms of power).

Losses and Efficiency

The basic definition of efficiency (the ratio of output divided by input) applies to all ma-

chines. The efficiency of a motor is significantly less than unity, so it is often possible
to obtain realistic results by directly measuring the input and output power under load
and calculating the efficiency. However, it is
sometimes easier to determine the losses and
then calculate the efficiency using Equation
1–14. What we will do here is list the losses,
point out how they are affected by loading
the motor, and briefly explain how to obtain
the required measurements and how to do
the calculations involved. We have listed the
losses for a long-shunt compound motor
equipped with interpoles and compensating
windings. For a short-shunt machine, use the
line current (shunt field current plus armature current) to find the series field losses, but
all other calculations are unchanged. If any
winding in our list does not exist on the machine, omit it from the calculations. The input power is always the product of line voltage and line current. Direct-current motor
losses are as follows:

Rotational Losses

■ Iron losses: hysteresis and eddy current
losses in the armature core and a small
amount of such losses in the pole faces.
■ Mechanical losses: that is, windage, and
friction loss at the brushes and in the
bearings.

In the absence of any better information,
consider the sum of these losses to be directly
proportional to the rotational speed of the
armature and (at the no-load speed) very
nearly equal to the armature power input at
no load (subtract the no-load armature circuit losses for more accurate results).

Shunt Field Copper Loss

The shunt field losses are equal to the
product of voltage and current at the shunt

field terminals. Do not use the line voltage for
this calculation unless it is desired to include
the field rheostat losses. If the field rheostat
is not changed, these losses remain constant.

Armature Circuit Losses

■ Series field loss
■ Compensating winding loss
■ Interpole winding loss
■ Armature copper loss
■ Brush contact loss

These losses are sometimes called variable
losses because they change if the motor load
changes. At no load, the armature current is
small, and these losses become very small.
That is why the no-load armature power input
is almost entirely rotational losses.

The armature circuit losses in the windings
are all proportional to the square of the armature current. If we know the armature current
and the resistance values, they can be easily
calculated using the I^2R approach. But the
brush contact loss does not vary in the same
manner. Instead, the voltage drop due to
brush contact resistance tends to be constant
at about 2 volts (V) no matter how much
current flows, and this voltage drop times the
armature current will give us the brush contact loss.

Since the armature circuit losses in an existing motor cannot be directly measured, we
need some way to calculate them. The simplest procedure is to determine an equivalent
armature circuit resistance[2] that will include
the effect of brush contact loss, and use I^2R.
For better accuracy we could determine the
resistance of windings[3] and assume 2-V brush

[2]Because of the variable nature of the brush contact
resistance, ohmmeter measurements are not satisfactory.
A voltmeter–ammeter technique using approximately
full-load current will give reasonable results.
[3]An ohmmeter of suitable range can be used for this
purpose.

drop. Better yet, Forgue's method[4] gives accurate results for both winding resistance and brush contact voltage drop.

As in most other machines, the losses produce heat in the associated parts, and the temperatures rise until the heat can be dissipated as fast as it is produced. Most motors are air cooled, having a fan to circulate air through the machine.

SOME ASPECTS OF ARMATURE WINDING DESIGN

Motors with Four or More Poles

It is quite common for dc motors to have more than two main field poles. Four, six, and eight pole designs are common, and any even number of poles is theoretically possible. In such machines, adjacent main poles always have opposite magnetic polarity. The flux lines go from the north field poles through the armature teeth, partway around the armature behind the slots, and back into the field structure at the south poles, as shown in Figure 2-6. If the motor has interpoles,

their number will either be the same as or else half of the number of main poles. But no matter how many poles the motor has, the armature core remains essentially the same. Only the armature winding is specifically designed to work with a given number of field poles.

Shims are sometimes found between the pole cores and yoke. The purpose of these is to adjust the air gaps between the poles and the armature so that the main poles will have equal flux densities and the interpoles will be the correct strength relative to the armature mmf. If a field structure is to be disassembled, mark the location of every pole core and every shim so that they can be reinstalled correctly.

[4]This procedure can be found in Donald V. Richardson, *Rotating Electric Machinery and Transformer Technology* (Reston, Va.: Reston Publishing Company, Inc., 1978). Richardson treats the laboratory testing of most electrical machines in considerable detail.

FIGURE 2-6 Typical magnetic circuits for four- and six-pole motors

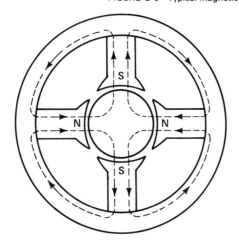
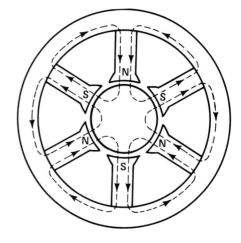

The required spatial distribution of current in motors with four or more poles is completely analogous to that of a two-pole machine. All the armature conductors in front of the north pole faces must carry current one way and those in front of the south poles must carry current the other way, as shown in Figure 2-6. A neutral plane necessarily exists about halfway between every pair of adjacent field poles, and this is where commutation must occur.

It is common practice to arbitrarily define the angular space between the center lines of adjacent main field poles as 180 electrical degrees. This is done because most important angular displacements in a machine are the same when expressed in electrical degrees. For example, the distance between the center line of a main pole and that of the nearest interpole is always 90° electrical no matter how many poles there are in the motor. The distance between two most nearly adjacent main field poles of like magnetic polarity is always 360° electrical. We can use electrical degrees to express angular distance around an armature core or a commutator as well, but sometimes it is more convenient to express these angles as so many slots or so many commutator bars. Equation 2–4 shows how to convert angles from one terminology to the other. The words "span" or "pitch" are frequently used to indicate an angular distance.

$$\phi = \frac{180PX}{n} \text{ (electrical degrees)} \quad (2\text{–}4)$$

where ϕ = angular displacement, in electrical degrees

X = angular displacement, in slots or bars

P = number of poles

n = total number of slots or bars

The distinctive difference between armature windings designed for various numbers of poles is the coil span or coil pitch, that is, the distance between the two sides of any coil. In a two-pole motor, the coil span is nearly 180° mechanical, but in four-pole machines it is about 90° mechanical, and 60° mechanical in a six-pole unit, that is, about 180° electrical in all cases. If the coil span is exactly 180° electrical, we say it is a full pitch winding. However, coil spans of less than 180° are common, and such windings are described as chorded or fractional pitch windings. A modest reduction of coil span (down to 150° or so) has several advantages (e.g., it reduces the length of wire in the coils and therefore reduces armature resistance) and does not adversely affect machine performance.

When the coils are connected to the commutator, the space (i.e., the number of commutator bars) between the two ends of any one element is known as the commutator pitch. It is the commutator pitch that defines whether an armature is lap or wave connected, and it also determines the degree of multiplicity and the degree of reentrancy.

Lap windings always have rather small commutator pitch. In their simplest form, lap windings have a commutator pitch of 1, and have as many parallel paths as there are poles.

Wave windings have a commutator pitch that is approximately (but never exactly) 360° electrical. In their simplest form, wave windings have a commutator pitch as near as possible to 360° and have only two parallel paths, regardless of the number of poles. It is interesting to note that if we attempt a wave connection on a two-pole armature, we find it awkward to do, and the result is not fundamentally different from a lap connection. Two-pole wave windings are therefore nonexistent.

It is possible to change the commutator pitch and end up with two, three, or four times as many parallel paths in the winding (compared to the simplest form). Whether lap or wave, the simplest windings are called simplex windings and have the smallest possible number of parallel paths. Duplex, triplex, and quadruple windings have two, three, or four times as many parallel paths as simplex windings, and we say their degree of multiplicity is two, three, or four, respectively. The multiplicity of simplex windings is 1. From a design point of view, the possibility of changing the number of parallel paths is important, sometimes because it enables us to circumvent commutation problems, and sometimes because changing the number of paths permits changes in the number of turns and size of wire in the coils, and this may permit a more economical design.

The term "reentrancy" refers to the number of closed loops formed by the completed armature winding. An armature winding always forms at least one closed loop (described as singly reentrant), but it may form two, three, or four loops that are electrically isolated from each other. Such windings are said to be doubly reentrant, triply reentrant, or quadruply reentrant. This is usually not of much concern to a designer, but it can affect the results of certain repair-shop tests. Reentrancy is always equal to the highest common factor between the number of elements and the degree of multiplicity.

The number of sets of brushes is normally equal to the number of field poles. Nonequalized lap and frog-leg windings require this arrangement, and adjacent sets of brushes always have opposite electrical polarity. However, wave windings and fully equalized lap windings can be operated with only two sets of brushes 180° electrical apart.

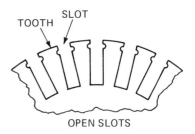

OPEN SLOTS

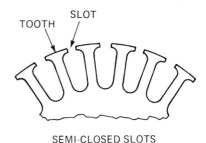

SEMI-CLOSED SLOTS

FIGURE 2-7   Open and semi-closed slots in an armature core

Slot Shapes and Conductor Cross Sections

Small armatures generally use semiclosed slots, as shown in Figure 2-7. In such a case the coils are wound into the slots and connected to the commutator. The whole assembly is balanced mechanically and then dipped in insulating varnish and baked, thus forming a rigid mass. The wire used for the winding is usually of circular cross section and the largest size that space will permit.

Larger armatures often have open slots, as shown in Figure 2-7. In this case the wires are usually of square or rectangular cross section, and the coils are wound on a form, bent to shape, wrapped with insulation, dipped in varnish and baked; they are then placed in the slots and connected to the commutator.

Semiclosed slots allow the field flux to be

more uniformly distributed along the air gap and so reduce the reluctance of the magnetic circuit. They also inherently provide better support for the coils to prevent them from flying out of the slots owing to centrifugal force. However, round wires do not pack very well, and there is considerable wasted space in each slot. The square or rectangular wire used with open slots inherently makes better use of the available space, but additional mechanical support for the coils (in the form of steel or phosphor bronze banding wires or possibly fiber-glass bands) will be required.

Factors That Affect Reactance Voltage

Since excessive reactance voltage causes commutation problems, it is desirable to keep it as low as possible. There are three design alterations that will reduce the reactance voltage. First, avoid 180° coil spans so that elements which commutate at the same time will not share a common slot. This reduces the mutual inductance between those elements and therefore reduces the reactance voltage. Second, use more elements and more commutator bars and reduce the number of turns per element so that the inductance of each one is reduced. Third, design the winding with more parallel paths so that the current in each individual element will be reduced. The usual maximum is about 300 A per path.

Numbers of Slots, Elements, Coils, and Commutator Bars

Most manufacturers produce motors having a variety of speed and power output ratings. However, for economic reasons it is desirable to minimize the number of different parts

that must be obtained. It is quite common to find two or more different armatures that are built using exactly the same punchings or laminations, and sometimes even having identical commutators. A designer must therefore know what constraints are imposed by having a fixed number of slots in the core and/or a fixed number of commutator bars. Rewinding an armature presents a similar problem. The new winding must be an exact duplicate of the old, and so one of the first steps is to determine exactly how the original was done. If the person doing the work understands the design constraints, there is much less probability of making a mistake. We will therefore examine the relationships between the numbers of slots, elements, coils, and commutator bars.

First, we must consider the number of commutator bars that could theoretically be used. Lap windings impose no constraints, but wave and frog-leg windings do and those restrictions are inviolate. However, for a winding that requires, say, 47 commutator bars, manufacturers have been known to use a commutator with 48 bars and jumper two adjacent bars together, thus reducing it to 47 effective bars. This particular arrangement is not very common, so from here on we will assume that there is no difference between the actual and effective number of bars.

On some armatures the number of slots is equal to the number of commutator bars, but this is not always the case. From a design standpoint, a large number of slots means narrow teeth, and if the teeth become too narrow, centrifugal forces may cause them to break. So the number of slots is usually much lower than the number of commutator bars. The preferred arrangement is that the number of slots shall be an integral dividend or factor of the number of commutator bars. If the number of bars is two, three, or four times the number of slots, we can wind the coils with two, three, or four strands of insulated wire (so that they remain isolated from each other), and then treat each strand as a separate element that is connected to its own commutator bars. If this is done, we call each strand an element, and the group of strands that have been wound together is known as a multielement[5] coil. A coil wound with only one strand is a single-element coil.

Another inviolate rule of lap or wave windings is that the number of effective elements must always equal the number of commutator bars. To maintain mechanical balance, all coils on the armature are generally the same, and therefore the actual number of elements will be an integral multiple of the number of slots. If the actual number of elements exceeds the number of bars, some dead elements must exist in that winding. A dead element is one that is not connected to the commutator. If there is more than one, they will be equally spaced around the armature periphery.

Lap Windings

No matter how many poles or what the degree of multiplicity may be, a lap winding can be done with any number of elements and commutator bars. The multiplicity is always equal to the commutator pitch, and the number of parallel paths in the winding is always equal to the product of the number of poles and the degree of multiplicity.

There are two ways of drawing armature windings. One way is to picture it laid out on a flat surface, as shown in Figure 2-8. Note that we have numbered the slots and the commutator bars, and that some of these

[5]If the various strands of a coil are all connected to the same pair of commutator bars, they are just parallel strands in a single-element coil.

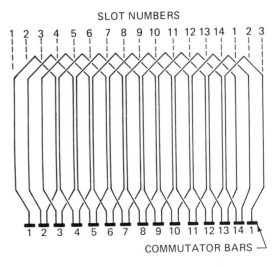

FIGURE 2-8 Four-pole lap-wound armature

are shown twice (once on the right and once on the left end of the diagram). We also show the elements as having only one turn, but they could have almost any number.

The other style of armature winding diagram is shown in Figure 2-9. It is drawn as the winding would appear viewed from the commutator end, but to show the connections clearly the other ends of the coils are shown "flared out" and the brushes and commutator appear "inside out."

The windings in Figures 2-8 and 2-9 are both the same. They represent a four-pole simplex lap-wound armature that has 14 single-element coils and 14 commutator bars. The coil span is three slots, the commutator pitch is one bar, and the degree of reentrancy is 1. All this information can be seen in either diagram.

Figure 2-10 shows a four-pole simplex lap-wound armature with 9 slots, 9 dual-element coils, and 18 commutator bars. The coil pitch is 2, the commutator pitch is 1, and the degree of reentrancy is 1.

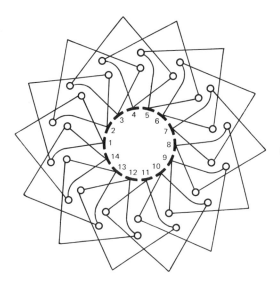

FIGURE 2-9   Four-pole lap-wound armature

FIGURE 2-11   Duplex lap-wound armature with
nine slots

Figure 2-11 shows a four-pole duplex lap-wound armature with 9 slots, 9 dual-element coils, and 18 commutator bars. The coil span

FIGURE 2-10   Simplex lap-wound armature with
nine slots

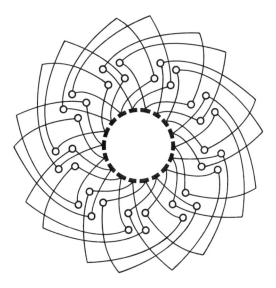

is 2, the commutator pitch is 2, and it is a doubly reentrant winding.

In the armature winding diagrams in Figures 2-8 through 2-11 the ends of each element come "straight down" to their respective commutator bars. This is known as a progressive connection. If the ends of each element cross as they go to the commutator, the winding is said to be retrogressively connected. Figure 2-12 illustrates this difference.

Lap windings inherently have more parallel paths than wave windings and so are more suitable for motors of high current ratings. However, the cemf in any one path is generated by conductors under only two of the field poles. If the field poles are not of equal strength (e.g., if the armature is not properly centered), the current flow in the various paths will not be equal. This gives rise to additional heating in the coils and also causes poor commutation. Equalizer connections will help to overcome the commutation problem.

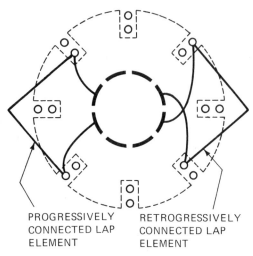

PROGRESSIVELY   RETROGRESSIVELY
CONNECTED LAP   CONNECTED LAP
ELEMENT   ELEMENT

FIGURE 2-12 Progressive and retrogressive
connections

On armatures with more than two poles, the subsequent pole pairs are just a repetition of the first two. If the number of elements is an integral multiple of the number of poles, there must be several groups of $P/2$ elements that are at exactly the same potential. If we take the number of elements and divide by half the number of poles, the resulting quotient is the number of such groups of equipotential elements in the armature winding, and the quotient tells us the spacing of those elements. For the armature in Figure 2-8 or 2-9, the quotient is 7, so there are 7 sets of equipotential elements, that is, element pairs 1 and 8, 2 and 9, 3 and 10, and so on, down to 7 and 14. For a six-pole armature with 54 elements, there will be 18 such groups: $1 + 19 + 37$; $2 + 20 + 38$; $3 + 21 + 39$, and so on, down to $18 + 36 + 54$. If the number of elements is not an integral multiple of half the number of poles (so that our quotient is not an integer), equipotential elements do not exist and the winding cannot be equalized.

There are two ways of doing equalizer connections. One way is to put jumpers between the appropriate commutator bars. If this is done we may say the armature has a cross-connected commutator. In such a case it is usual to put in all the possible cross connections, and we say that the armature is fully equalized. But if only half the possible equalizing connections are installed, we say that it is only 50% equalized. For a partially equalized armature, the connections are installed as symmetrically as possible.

If the armature has only single turn elements, it may be easier to install equalizer connections at the end opposite the commutator. The connections can quite easily be made at the extreme end (knuckle) of the coil, but space limitations usually make 100% equalization impractical.

Wave Windings

If you start at one commutator bar and trace the circuit of a wave winding through $P/2$ elements, you will arrive at a bar that is very near your starting point. Doing this tracing operation on a simplex wave winding will bring you back to the bar adjacent to your starting point, but in duplex, triplex, or quadruplex types, you end up two, three, or four bars from your starting point. It therefore follows that wave windings are restricted to certain numbers of coils and bars. The relationship between the number of bars and the commutator pitch (Y_c) is given by Equation 2-5, and the only restriction is that Y_c must be an integer.

$$Y_c = \frac{C \pm m}{P/2} \qquad (2-5)$$

where C = number of commutator bars
 m = degree of multiplicity

If we apply Equation 2-5 to a four-pole simplex wave armature with 13 commutator bars,

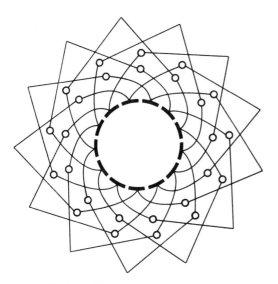

FIGURE 2-13 Retrogressive wave winding

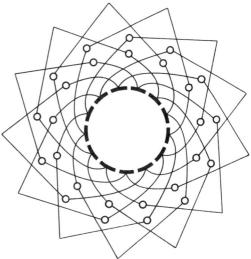

FIGURE 2-14 Progressive wave winding

we find that Y_c must be either 6 or 7. If we use 6, the winding ends up as shown in Figure 2-13. Since tracing through $P/2$ elements brings us out one commutator bar behind our starting point, we call this a retrogressive winding. If we use 7, the winding will be as shown in Figure 2-14. In this case, tracing through $P/2$ elements will put us one bar ahead of our starting point (in the direction of our tracing operation), and so it is called a progressive winding. Both of the windings shown have a coil span of 3 and are singly reentrant.

EXAMPLE 2-1

It is desired to build a ten-pole triplex wave armature with between 200 and 220 commutator bars. What numbers of commutator bars are possible?

Solution

From Equation 2–5,

$$Y_c = \frac{C \pm 3}{5}$$

If we use (−3) in the formula, the smallest pos-sible value for c (for which Y_c is an integer) is 203, and that would make $Y_c = 40$. If Y_c goes to 41, C then becomes 208, and so on. Similarly, if we use (+3), the smallest possible value for C is 202, for which $Y_c = 41$. If we now increment the value for Y_c, we get the following values for C: 202, 203, 207, 208, 212, 213, 217, 218.

The number of parallel paths in a wave winding is always twice the degree of multi-plicity. Since each parallel path contains conductors under every field pole, nonuni-formity of field pole strength does not cause heating or commutation problems.

Frog-Leg Windings

A frog-leg winding is a combination of a lap and a wave winding connected to the same commutator. It generally results in better commutation than either lap or highly multi-plexed wave windings.

Considering any two commutator bars, the instantaneous voltage produced by the lap elements in series between those bars must equal the instantaneous voltage of the wave elements connected between those bars. Otherwise, circulating currents will be set up, causing objectionable heating. The easiest way to ensure the equality is to wind the lap and wave elements with the same number of turns and the same coil span.[6] The number of

lap elements and the number of wave elements must each equal the number of commutator bars, and so the problem is reduced to that of making the connections to the commutator, with due regard to the total voltage, relative phasing, and polarity of the lap and wave elements.

To make the lap and wave voltages come out equal, the two windings must have the same number of parallel paths. If the lap section is simplex (and it usually is), the multi-

[6]This is possible if the number of slots is an integral multiple of the number of poles. But if the number of slots is an integral multiple of only half the number of poles, the span of the lap and wave coils must be different and such that the sum of their spans is equal to 360° electrical.

FIGURE 2-15   Simple frog-leg winding

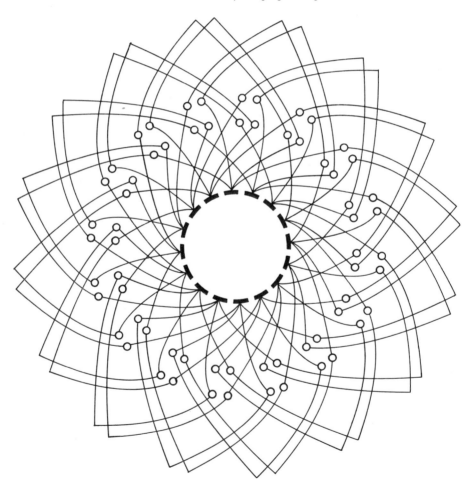

plicity of the wave section must equal $P/2$. The commutator pitch of the wave winding then becomes 360° electrical plus or minus one bar.

The get the relative polarity the same, one section of the winding (either the lap or the wave elements) must be progressively con-nected while the other must be retrogres-sively connected.

Figure 2-15 shows a frog-leg winding done with 16 coils, 32 elements, and 16 commuta-tor bars. The coil span is 3, and the commu-tator pitches are 7 and 1 for the wave and lap coils, respectively.

MATHEMATICAL INTRODUCTION TO DC MOTORS

Torque in a DC Motor

Equation 1–10 is valid for a dc motor and yields developed (not output) torque.

EXAMPLE 2-2
A six-pole simplex lap-wound armature with 240 three-turn elements has a diameter of 0.35 m and its length is 0.9 m. If the main field poles cover 65% of the armature periphery and the average flux density under those poles is 0.93 Wb/m², find the torque developed when the total armature current is 90 A.

Solution
A six-pole simplex lap winding has six parallel paths; therefore,

$$I = \frac{90}{6} = 15 \text{ A}$$

From Equations 1–8 and 1–9,

$$n = N_A \times 0.65$$
$$= 2 \times 240 \times 3 \times 0.65$$
$$= 936 \text{ conductors}$$
$$T_D = \frac{0.93 \times 0.9 \times 15 \times 936 \times 0.35}{2}$$
$$= 2056.5 \text{ N-m} \qquad \text{(Answer)}$$

Counter EMF Calculations

Equation 1–7 is valid for motors as well as generators.

EXAMPLE 2-3
If the armature from Example 2-2 is rotating at 3.6 rev/s, find the cemf.

Solution
Since the critical specifications of the two ar-matures are the same, see Example 1-1.

$$\text{cemf} = 517 \text{ V} \qquad \text{(Answer)}$$

Armature Current Calculations

If we use the concept of effective armature resistance given on pp. 34–36, then Equation 2–1 should be used. If we wish to keep the contact voltage drop separate from the actual winding resistances we use Equation 2–6.

$$I_A = \frac{V - V_B - E_G}{R_{AC}}$$

where V = applied voltage
V_B = brush contact drop
E_G = cemf
R_{AC} = armature circuit resistance, not including brush contact resis-tance

EXAMPLE 2-4
Referring to the armature in Example 2-2, as-suming a 600-V power supply and constant field flux, find the following:

a. The effective armature circuit resistance (R_{AE}).

b. Starting current (if starting resistance is not added).
c. The starting current if 5.0-ohm (Ω) resistance is added to the armature circuit.

Solution

a. $I_A = \dfrac{V - E_G}{R_{AE}}$

$90 = \dfrac{600 - 517}{R_{AE}}$

$R_{AE} = \dfrac{83}{90}$

$= 0.9222 \ \Omega$ (Answer)

b. $I_A = \dfrac{600}{0.9222}$

$= 650.6 \text{ A}$ (Answer)

c. $I_A = \dfrac{600}{5.9222}$

$= 101.3 \text{ A}$ (Answer)

EXAMPLE 2-5

Referring to the armature in Examples 2-2 and 2-3, assuming a 600-V power supply, constant field flux, and 2-V brush contact drop, find the following:

a. The armature circuit resistance excluding brush contact drop (R_{AC}).
b. The starting current (if starting resistance is not added).
c. The starting current if 5.0-Ω resistance is added to the armature circuit.

Solution

a. $I_A = \dfrac{V - V_B - E_G}{R_{AC}}$

$90 = \dfrac{600 - 2 - 517}{R_{AC}}$

$R_{AC} = \dfrac{81}{90}$

$= 0.9 \ \Omega$ (Answer)

b. $I_A = \dfrac{600 - 2 - 0}{0.9}$

$= 664.4 \text{ A}$ (Answer)

c. $I_A = \dfrac{600 - 2 - 0}{5.9}$

$= 101.4 \text{ A}$ (Answer)

Input, Output, Losses, Efficiency, and Output Torque

The basic relationships between input, output, losses, and efficiency are unchanged. However, do not forget that motor armature current equals the line current minus the shunt field current. Remember, too, that losses in a motor decrease the output torque available. At no load the input power is just whatever losses exist. If we have sufficient information, we can consider the no-load armature circuit losses separately. It is also useful to note that the product of E_G and I_A is always equal to the developed power.

EXAMPLE 2-6

Referring to the motor from Examples 2-2 through 2-5, assume 750-W rotational loss. Then find the following:

a. The developed power and compare it to the product of E_G and I_A.
b. The output power.
c. The output torque.
d. The efficiency (assume the field copper loss to be zero).
e. The efficiency if the motor has a shunt field that draws 3 kW.

Solution

a. $P_D = 2\pi T \text{ (rev/s)}$

$= 2 \times 3.1416 \times 2056.5 \times 3.6$

$= 46{,}517 \text{ W}$

$= 46.517 \text{ kW}$ (Answer)

$E_G I_A = 517 \times 90 = 46{,}530$ W

$\cong$ developed power

b. Output power $= 46{,}530 - 750$

$= 45{,}780$ W (Answer)

c. $45{,}780 = 2 \times 3.1416 \times T_o \times 3.6$

$T_o = 2023.9$ N-m (Answer)

d. $n = \dfrac{\text{output}}{\text{input}}$

$= \dfrac{45{,}780}{600 \times 90}$

$= 0.8478$ (Answer)

e. $n = \dfrac{45{,}780}{(600 \times 90) + 3000}$

$= 0.80316$ (Answer)

Calculation of Speed and Speed Regulation

If the armature winding data, the field flux, and the applied voltage are known, it is easy to calculate the speed of a motor for any value of armature current. Such a problem is basically the reverse of Example 2-3.

If the developed torque is given or can be found, the armature current can be found using the reverse of Example 2-3. But if the load is given in kilowatts, we are forced to either do an approximation or solve with a quadratic equation, as shown in Example 2-7.

EXAMPLE 2-7

Referring to the armature from Examples 2-1 through 2-5, find the following:

a. The cemf using an approximation and based on equivalent armature resistance.
b. The cemf using a quadratic equation, assuming 2-V brush drop and using R_{AC}.
c. The no-load speed using the answer for part a.
d. The no-load speed using the answer for part b.

e. The speed regulation based upon your answer for part c.
f. The speed regulation based upon your answer for part d.

Solution

a. The developed power is always equal to the sum of rotational losses plus output power. In this case the output is zero, so developed power becomes 0.75 kW. At no load the cemf will be nearly 600 V. Using that value,

$$I_A = \frac{750}{600} = 1.25 \text{ A}$$

and so a better value for cemf is obtained using Equation 2-1.

$$1.25 = \frac{600 - E_G}{0.95}$$

$$\therefore E_G = 600 - (1.25 \times 0.9222)$$

$$= 598.85 \text{ V} \qquad \text{(Answer)}$$

b. From Equation 2-6,

$$I_A = \frac{600 - 2 - E_G}{R_{AC}}$$

$$\therefore E_G = 600 - 2 - I_A R_{AC}$$

since

$$P_D = E_G I_A$$

$$I_A = \frac{P_D}{E_G}$$

Substituting

$$E_G = 600 - 2 - \frac{P_D}{E_G}$$

$$E_G^2 = 598 E_G - P_D$$

$$E_G^2 - 598 E_G + 750 = 0$$

$$E_G = \frac{598 \pm \sqrt{598^2 - 4(750)}}{2}$$

$$= 596.74 \text{ V} \qquad \text{(Answer)}$$

or

$= 1.2568$ V (we will discard the figure 1.2568 V as impractical)

c. $E_G = \dfrac{P\phi N_A}{6}$ (rev/s)

$598.85 = \dfrac{0.5982 \times (2 \times 240 + 3)}{6}$

$\times$ (rev/s)

rev/s = 4.1712   (Answer)

d. $E_G = \dfrac{P\phi N_A \text{ (rev/s)}}{6}$

$596.74 = \dfrac{0.5982 \times 2 \times 240 \times 3 \text{ (rev/s)}}{6}$

rev/s = 4.1565   (Answer)

e. Speed regulation = $\dfrac{4.1712 - 3.6}{3.6}$

= 0.1587   (Answer)

f. Speed regulation = $\dfrac{4.1565 - 3.6}{3.6}$

= 0.1546   (Answer)

SUMMARY

The construction of a dc motor is virtually identical to that of a generator. The theory of operation is also substantially the same. However, for a generator we tend to regard its terminal voltage as a dependent variable, and for a motor we usually consider the speed to be the dependent variable. Although speed regulation of a motor is essentially the same phenomena as voltage regulation of a generator, the change of emphasis requires some realignment of our thinking.

Because of the way they are used, there are some things about generators that simply do not apply to motors. In motors there is nothing analogous to the use of equalizer connections for parallel operation and nothing comparable to the voltage buildup process of self-excited generators. But the similarities are many.

The most complex part of a dc motor is probably the armature. But armature windings follow a rather few distinct patterns, and once a specific pattern is recognized it becomes fairly easy to follow on a diagram and/or to duplicate in a repair shop. Although we did not examine armature windings in Chapter 1, generator armature windings are the same.

The content of this chapter should be regarded as preparatory material. Readers who are interested in the application of or troubleshooting dc motors should study Chapters 8 through 10. Many aspects of dc motor repair and/or design have been omitted from this book, and students who are interested in these topics must look to other sources of information.

QUESTIONS

2-1. Explain what is meant by the following terms:
      (a) Counter electromotive force
      (b) Starting torque
      (c) Starting current
      (d) Shunt (adjective)

(e) Run away (action of a motor)
(f) Compound
(g) Torque
(h) Developed power
(i) Semiclosed slots
(j) Coil span
(k) Commutator pitch
(l) Wave winding
(m) Lap winding
(n) Multiplicity
(o) Reentrancy
(p) Reactance voltage
(q) Armature coil
(r) Armature element

2-2. The line current and armature current in a motor are reversed compared to a generator. Why?

2-3. The reversed armature current causes five observable differences between generators and motors. What are the five differences?

2-4. A permanent magnet motor tends to be more efficient than the other types. Why?

2-5. Some motors designed for separate excitation can be operated as shunt motors while others cannot. Explain why?

2-6. What causes a motor to run away?

2-7. Draw schematic diagrams of a long-shunt and a short-shunt motor.

2-8. How will the speed and/or direction of rotation of a motor be affected by the following changes (assume all other factors remain constant):
(a) Armature voltage is reduced.
(b) Armature voltage is reversed.
(c) Armature circuit resistance is increased.
(d) Field flux is reduced to zero.
(e) Field flux is increased.
(f) The load torque reverses (so that it tends to push the motor).

2-9. What are the three basic methods of obtaining speed control?

2-10. For a continuous low-speed operation, armature resistance control is the least desirable speed-control method. Why?

2-11. Considering an elementary shunt-wound motor, which part (the field coils or the armature) will overheat if
(a) The motor is overloaded?
(b) The supply voltage is much too high?
(c) The supply voltage is much too low and the load torque remains at the full-load value?

2-12. If the brushes are moved away from their correct position, the motor torque per ampere will decrease. Explain why.

2-13. Why does moving the brushes away from their correct position raise the motor speed?

2-14. Why must the location of each pole core and each shim be marked when a field structure is disassembled?

2-15. How can one recognize the difference between an eight-pole field structure with no interpoles and a four-pole field that has four interpoles?

2-16. There must be at least two open circuits in an armature winding before either of them can be found by doing a continuity test between commutator bars. Explain why.

2-17. Given an armature that has 144 commutator bars, explain how to determine whether it is a simplex, duplex, triplex, or quadruplex winding by using a continuity test device.

2-18. Draw a diagram of a complete motor in the following three steps:
(a) Draw a diagram (similar to Figure 2-8) of a two-pole simplex lap-wound armature that has six slots

in the core, six elements, and six commutator bars.

(b) Show the brushes, assign them an electrical polarity, and show the current direction in each conductor.

(c) Draw field poles in the correct relative position, assign them a magnetic polarity, and determine the direction of rotation.

2-19. What is the advantage of rectangular or square wire for armature windings?

2-20. What three things can a motor designer do to minimize reactance voltage?

2-21. A motor that has 32 slots and 60 commutator bars must have 4 dead elements. Explain why.

2-22. Draw a diagram of a four-pole simplex lap-wound armature that has 8 slots, 8 two-element coils, and 16 commutator bars. Use a coil span of 180° electrical and specify it in slots.

2-23. What is the main purpose of equalizer connections and in what two ways may they be made on an armature?

2-24. Draw a diagram of a four-pole simplex wave-wound armature that has 8 slots, 8 dual-element coils, and 15 commutator bars. Use full-pitch coils.

2-25. In the diagram for question 2-24.
(a) How many dead elements appear?
(b) Is the winding progressively or retrogressively connected?

2-26. Draw a diagram of a frog-leg winding that has 14 slots, 14 dual-element coils, and 14 commutator bars. Use a coil span of 3 for the wave elements and 4 for the lap elements.

MATHEMATICAL PROBLEMS

2-1. If a motor is developing 75 N-m of torque, find its torque if
(a) The armature current rises to 160% of its original value.
(b) The field flux drops to 90% of its original value.
(c) Both of the preceding changes occur.

2-2. If the cemf of a certain motor is 105 V when operating at 22.5 rev/s, find its cemf when
(a) The speed rises to 29.2 rev/s with no change of field flux.
(b) The field flux increases 25% with no change of speed.
(c) The speed rises to 29.2 rev/s and the field flux drops to 60% of its original value.

2-3. If the applied voltage is 125 V and the armature resistance is 1.3 Ω, find the armature current when
(a) The cemf is zero.
(b) The speed is zero.
(c) The cemf is 95 V.
(d) The cemf is 110 V.

2-4. Find the motor efficiency, power input, power output, and/or total losses as appropriate for each case:
(a) Total power input = 700 W, power output = 500 W.
(b) Total power input = 25 kW, total losses = 4 kW.
(c) Line current = 25 A, line voltage = 125 V, output = 2.5 kW.
(d) V = 250 V, I_A = 20 A, shunt field current = 2 A, output = 3.2 kW.
(e) Power output = 15 kW, efficiency = 85%.

(f) Power input = 150 kW, efficiency = 91%.

(g) Rotational losses = 2 kW, shunt field losses = 3.2 kW, armature circuit losses = 7 kW total, output = 90 kW.

(h) Rotational losses = 300 W, shunt field current = 1.2 A, shunt field resistance = 87 Ω, effective armature circuit resistance = 0.9 Ω, armature current = 15 A, line voltage = 125 V.

(i) Rotational losses = 600 W, shunt field current = 2.0 A, shunt field resistance = 80 Ω, armature resistance = 1.1 Ω, armature current = 19 A, interpole winding resistance = 0.3 Ω, series field resistance = 0.45 Ω, long-shunt connection, line voltage = 250 V, brush contact drop = 2 V.

2-5. For each of the following armatures, find the number of degrees per slot (or per commutator bar) and the number of slots or bars in 180° electrical.
(a) 6 pole, 72 slots
(b) 8 pole, 240 bars
(c) 4 pole, 28 slots
(d) 14 pole, 96 slots
(e) 4 pole, 49 bars

2-6. Find the number of elements per coil and the number of dead elements in each of the following windings.
(a) 28 bars, 14 slots, lap wound
(b) 54 bars, 18 slots, lap wound
(c) 71 bars, 24 slots, wave wound
(d) 96 bars, 48 slots, frog-leg winding
(e) 140 bars, 72 slots, frog-leg winding
(f) 106 bars, 54 slots, wave wound

2-7. For each of the armatures in problem 2-5, find the largest possible coil span that is less than 180° electrical.

2-8. For each of the following wave-wound armatures, find the commutator pitch or possible pitches:
(a) 71 bars, 4-pole simplex
(b) 70 bars, 4-pole duplex
(c) 94 bars, 6-pole duplex
(d) 96 bars, 6-pole triplex

2-9. For each of your answers for problem 2-8, specify whether the winding will be progressive or retrogressive and specify the degree of reentrancy.

2-10. Given an armature with 144 commutator bars, determine whether it can be frog-leg wound for 2, 4, 6, 8, 10, or 12 poles, respectively, and find the possible commutator pitches for the wave elements on those windings that can be done (assume the lap elements have a commutator pitch of 1).

2-11. Find the torque and cemf developed by each of the following armatures:
(a) Length of core = 0.3 m, diameter = 0.14 m; 2-pole simplex lap winding, total armature current = 50 A, 28 ten-turn elements, flux density = 0.8 Wb/m^2, pole coverage = 60%, 29 rev/s.
(b) Length of core = 2.0 m, diameter = 0.9 m, 6-pole triplex wave winding, total armature current = 1100 A, 178 single-turn elements, flux density = 0.75 teslas, pole coverage = 70%, 3.33 rev/s.

2-12. If their effective resistances are 1.656 Ω and 0.08791 Ω, respectively, find the voltage applied to each armature in problem 2-11.

2-13. Find the full-load torque of the following motors:
(a) 50 kW, 19 rev/s
(b) 0.2 kW, 59 rev/s
(c) 1000 kW, 3.75 rev/s

2-14. If the rotational losses of a motor are 1.5 kW at 14.6 rev/s,
    (a) How much torque is required to overcome these losses?
    (b) Find the rotational losses at 10 rev/s.

2-15. If a motor runs 19 rev/s at full load and 26.3 rev/s at no load, find its speed regulation.

2-16. Find the speed regulation and full-load efficiency of each of the following long-shunt motors.
    (a) Line voltage = 125 V, armature current = 85 A at full load, rotational losses = 700 W, effective armature resistance = 0.095 Ω, shunt field resistance = 19 Ω, field rheostat resistance = 6 Ω.
    (b) Line voltage = 250 V, I_A = 140 A at full load and 9 A at no load, shunt field resistance = 21 Ω, field rheostat resistance = 9 Ω, brush contact drop = 2 V, armature (only) resistance = 0.07 Ω, interpole coil resistance = 0.02 Ω, compensating winding resistance = 0.04 Ω, full-load speed = 11.5 rev/s.

TRANSFORMERS

Every industrially developed country in the world is overlaid by an electrical transmission and distribution network, the function of which is to transmit energy from the locations where it is readily available to the multitude of places where it is to be used. The devices that receive the energy are known as load devices or simply loads and are large in number, but most of them are comparatively small in size. Because of their modest individual power requirements it is not practical to design the loads to operate at voltages exceeding about 750 volts. In most countries, the loads (predominantly lighting equipment, heating devices, and motors) are scattered over a rather wide geographic area, but the most economical sources of energy tend to be found in a relatively few locations and sometimes are rather remote from the loads that require the energy. This is, of course, the reason for the development of the electrical transmission systems, and the basic problem is to transmit the energy with an acceptable degree of efficiency. But line loss (the power required to circulate current through the connecting wires of the electrical system) is di-

rectly proportional to the length of the lines and so limits the maximum practical distance over which power can be transmitted.

A second fundamental requirement is that an acceptable voltage must be maintained at the terminals of each load device. On most power systems the electrical loads are switched on and off according to the user's requirements, and the total load therefore fluctuates from time to time. Quite definite daily, weekly, and seasonal variations can be seen. Because of line drop (i.e., the voltage required to circulate current through the connecting wires of a system), the voltage at the loads tends to decrease at high load periods and recover its normal value when the load demand decreases. Elementary circuit theory also shows that the voltage on a distribution system will inherently tend to be lower at points remote from the source. To ensure compatability and to facilitate their mass production, load devices and all the other components of an electrical system must be designed for only a few preferred standard voltages. Unfortunately, a load device generally will not function properly unless the voltage at its terminals is within about plus or minus 10% of that for which it is designed, and this also tends to set a limit on the distance over which energy can be electrically transmitted.

Since the maximum permissible line drop tends to be a fixed percentage of the system voltage, the obvious solution is to transmit at the highest practicable voltage, and this idea has another advantage. Since the power transmitted is equal to the product of $V \times I$, a given amount of power can be transmitted at higher voltage using less current, and this decreases both the line drop and the line loss. To state the constraints another way, if distance and conductor size remain unchanged, raising the voltage by a factor of 10 will permit the transmission of 10 times the power and still reduce the percentage loss by a factor of 10. But we cannot eliminate the restriction of having our load devices rated at only a few hundred volts, and so the only answer to this dilemma is to transmit power at high voltages over most of the required distance and then to reduce it to an appropriate value at some point near the load.

The problem of economically and efficiently changing a dc voltage from one value to another has traditionally been and still is rather formidable. But the transformer makes it very easy to change ac voltage levels; that is its fundamental purpose in the system, and that is why dc power systems are comparatively rare. Alternating current systems are subject to voltage drop caused by inductive reactance that does not affect dc current, and, therefore, at a given voltage alternating current does not transmit as well as dc (direct current), but the advantage of convenient high-voltage transmission usually outweighs this drawback. Recent developments have made high voltage dc transmission possible, but for economic reasons there are probably less than a hundred such installations presently in service in the world.

The ability to receive power at one voltage (or current) level and deliver it at some other voltage (or current) level is the most important feature of transformers, but they do have other advantages. They can be used to derive (or, if you wish, can be used to supply power to) circuits that are isolated from the main power supply, they can be used to obtain voltages that are phase shifted from the main power supply, and they can be used to control the voltage on a power system. They can also be used for impedance-matching purposes, although this is seldom required on power circuits.

In this chapter we will begin by explaining the basic theory of operation. Then we will describe some of the more common commercial units, pointing out some additional features of construction. Then we will proceed with a mathematical treatment, some of which is related to transformer application and some of which will introduce the reader to the basics of transformer design.

CONSTRUCTION AND THEORY OF OPERATION

Fundamental Parts of a Transformer

To discuss the theory of operation, a transformer may be visualized as in Figure 3-1.

The essential parts are as follows:

Primary Winding

The word "primary" indicates that this coil is energized from an external source.

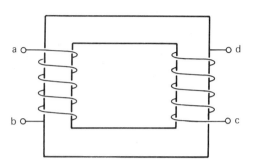

FIGURE 3-1   Elementary transformer core and coil assembly

Secondary Winding

The word "secondary" means that this coil is (or can be) used to deliver power to an external load.

Magnetic Circuit or Core

The main requirement for this part is that it must have high permeance. Its purpose is to provide such a good magnetic path that, when the primary coil is energized, all (or as many as possible) of the magnetic lines of force set up by the primary coil will link all the turns of the secondary coil. The third dimension is not shown, but the core has approximately a square cross section.

Theory of No-Load Operation

If the secondary winding is open circuited and an alternating voltage is applied to the primary winding, the current flow will be limited by the resistance and inductive reactance of the coil. If the resistance of the winding is small compared to its reactance (as it is in most transformers), the latter is then the most significant factor that limits current flow under these conditions. However, a coil has inductive reactance only because the magnetic flux it produces generates a voltage in the coil. This voltage is sometimes known as a self-induced voltage, but because

it opposes the applied voltage, it is also referred to as a counter emf or cemf. When dealing with transformer theory, it is most convenient to speak of cemf in the primary winding whenever we wish to designate the main current-limiting force that exists in that winding, and the term "reactance" (or "inductive reactance") will not be used until it has been given a unique definition.

Let us deal with applied voltage and cemf in graphical and then phasor terms. We will use the symbol v_1 to represent instantaneous cemf. The root-mean-square (rms) values will be represented by V_1 and E_1, respectively. We will furthermore add the subscripts a and b to indicate positive directions. Using these conventions, the equation $v_{1\,a-b} = 25$ volts (V) means that terminal a is 25 V positive with respect to b at this particular time. If $v_{1\,a-b} = -16$ V, then terminal a is 16 V negative with respect to b. If we write $E_{1\,b-a} = 14$ V, the instantaneous cemf is 14 V and opposing $v_{1\,a-b}$.

For graphs of voltages, positive values are shown above the zero line and negative values below. Graphs of $v_{1\,a-b}$ and $e_{1\,b-a}$ for the transformer in Figure 3-1 are shown as the smaller of the two, but on a power transformer they are almost equal and we often consider them to be so.

In similar fashion we will designate instantaneous primary current as i_1 and rms primary current as I_1, and the subscripts a and b will be used to indicate positive direction. Conventional current flow is assumed throughout. The equation $i_{1\,a-b} = 2$ A means the instantaneous current is 2 amperes and and flowing from a toward b. The current $i_{1\,a-b}$ has also been shown in Figure 3-2, and it lags $v_{1\,a-b}$ by nearly 90° of time.

The currents and voltages of Figure 3-2 are shown as phasors in Figure 3-3. Phasors rotate counterclockwise once for each cycle,

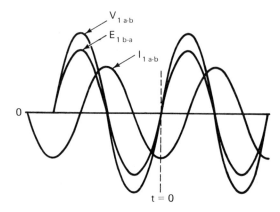

FIGURE 3-2   No-load primary voltages and current
in a transformer

and we simply stop the action at some convenient time and draw them in their correct positions. The relationship between a phasor diagram and a graph like that of Figure 3-2 is that, when the graph is positive, the phasor is in one of the first two quadrants. Figure 3-3 has therefore been drawn when $t = 0$ in Figure 3-2. If the core of the transformer has very high permeance, most of the flux set up by the primary will follow the core and link the secondary winding. It is convenient to think of each line of force as starting off as an infinitesimal little loop and "stretching out" to fit the core. Seen this way, one can visualize the flux as cutting across the conductors in the window of the core (i.e., the space enclosed by the magnetic circuit) and thus generating a voltage in each one. Since the same flux cuts each primary and each

FIGURE 3-3   No-load primary voltages and current
in a transformer

secondary conductor in the window, the voltage per turn must be the same in both windings, and the voltages must be in the same "wrap direction" around the core. Identifying the secondary voltage as E_2 and using subscripts c and d,

$$\frac{E_{1\,a-b}}{N_1} = \frac{E_{2\,c-d}}{N_2} \qquad (3\text{--}1)$$

where N_1 and N_2 are the numbers of turns in the primary and secondary windings, respectively.

In a practical sense, Equation 3–1 refers to internally developed voltages that cannot be easily measured with a voltmeter. However, the secondary terminal voltage $V_{2\,d-c}$ is the same as $E_{2\,c-d}$ (neglecting any loading effects of the meter), and $V_{1\,b-a}$ is approximately equal to $E_{1\,a-b}$, so we can rewrite the preceding equation as

$$\frac{V_{1\,b-a}}{N_1} \cong \frac{V_{2\,d-c}}{N_2} \qquad (3\text{--}2)$$

In the absence of any better information, Equation 3–2 is normally considered to be exact and is often given in terms of absolute rms values (i.e., meter readings) as shown in Equation 3–3.

$$\frac{|V_1|}{N_1} = \frac{|V_2|}{N_2} \qquad (3\text{--}3)$$

In words, Equation 3–3 says that the voltage per turn is the same in both windings. If we transpose Equation 3–3 to get Equation 3–4, the worded equivalent is "the voltage ratio equals the turns ratio."

$$\frac{|V_1|}{|V_2|} = \frac{N_1}{N_2} \qquad (3\text{--}4)$$

From Figure 3–2, one can recognize that any conductor that passes through the window of the core will have a voltage induced in it that will conform to Equations 3–1 through 3–4, and it would be quite possible to have a multitude of secondary windings. We shall consider some such arrangements later in the chapter.

Most power transformers are used with the supply system frequency held constant and the voltage nearly so. Assuming V_1 to be constant, the maximum flux in the core will be constant, and so the no-load current I_1 will be constant. The magnetizing ampere turns on the core $N_p I_1$ will therefore also remain constant.

Theory of Operation with Load

If we apply the appropriate right-hand rules to the transformer in Figure 3–4, we can see that when the applied voltage $v_{1\,a-b}$ is positive the primary winding will tend to magnetize the core in the direction shown, and the secondary voltage $e_{2\,d-c}$ will be positive. Since the secondary circuit is closed, the secondary current will tend to flow from terminal d to c inside the coil ($i_{2\,d-c}$ will be positive). It is immediately apparent that the secondary mmf will oppose the primary mmf (i.e., the secondary ampere turns tend to demagnetize the core). However, if the maximum flux set up in the core is reduced, the

cemf in the primary will be reduced, and because the winding resistance is small, even a small decrease of cemf will permit a large increase of primary current. So what happens is that the primary current increases enough to practically offset the mmf of the secondary, and the flux remains virtually unchanged.

To continue the argument mathematically, let us define I_o as the no-load primary current with positive direction understood to be from terminal a to b in Figure 3–4. If the flux in the core remains almost unchanged when load is applied, the sum of the ampere turns must remain almost constant, and we can write Equation 3–5 as shown.

$$N_1 I_{1\,a-b} + N_2 I_{2\,c-d} \cong N_1 I_o \quad (3\text{--}5)$$

Let us transpose Equation 3–5 as follows. First move the term containing I_2 over to the right-hand side, reverse the subscripts of I_2 and change its algebraic sign, and then divide through by N_1. The result is

$$I_{1\,a-b} \cong I_o + \frac{N_2}{N_1} I_{2\,d-c} \quad (3\text{--}6)$$

The right-hand term of Equation 3–6 is just the increase of primary current that occurs when a secondary load is connected to the transformer, and the phasor diagrams in Figure 3–5 are intended to illustrate this idea.

From Figure 3–5 it can be seen that if the no-load current of a transformer is small the primary current will lag (or lead) applied voltage by only a slightly larger (or smaller) angle than the angle between the voltage and current at the secondary load. Seen from the primary side and neglecting I_o, the transformer takes on the character of its secondary load, or, in other words, the power factor on the primary side is equal to the power factor of the secondary load. Mathematically,

FIGURE 3-4 Transformer under load

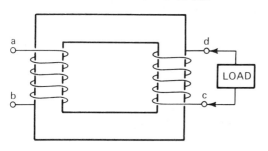

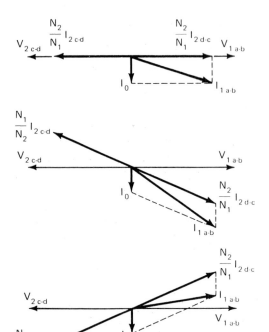

FIGURE 3-5 Phasor diagrams for a loaded
transformer: (top) unity power factor; (center)
lagging power factor; (bottom) leading power factor

$$\angle^{V_1} I_1 \cong \angle^{V_2} I_2 \qquad (3\text{-}7)$$

Let us reconsider Equation 3-5. If I_o is
negligible, the right-hand side becomes zero
and it can be transposed into

$$N_1 I_{1\,a-b} \cong N_2 I_{2\,d-c} \qquad (3\text{-}8)$$

It is convenient to rewrite Equation 3-8 in
terms of absolute rms currents as shown in
Equation 3-9. Basically, this says that the
ampere turns in the primary are practically
equal to the ampere turns of the secondary.

$$N_1 |I_1| \cong N_2 |I_2| \qquad (3\text{-}9)$$

This last equation can also be conveniently
written as shown in Equation 3-10, which

states that the voltage ratio is practically
equal to the inverse of the turns ratio.

$$\frac{|I_1|}{|I_2|} \cong \frac{N_2}{N_1} \qquad (3\text{-}10)$$

From Equations 3-4 and 3-10 we can derive
Equation 3-11, which states that the currents
are almost inversely proportional to the
voltages.

$$\frac{|V_1|}{|V_2|} \cong \frac{|I_2|}{|I_1|} \qquad (3\text{-}11)$$

This conveniently transposes as shown in
Equation 3-12, which shows that the apparent power input is practically equal to the
apparent power output.

$$|V_1||I_1| \cong |V_2||I_2| \qquad (3\text{-}12)$$

In the absence of more accurate information,
Equations 3-7 and 3-9 through 3-12 are taken
to be exact. When we do this, we are regarding the transformation as being ideal or
theoretically perfect, and large transformers
quite closely approach this performance.

Losses, Ratings, Cooling Methods, and Typical Efficiency

The real power input to a transformer is not
all delivered to the secondary load. Some of
this power is wasted simply pushing current
through the resistance of the windings, and
the alternating magnetic field creates a power
loss in the core as well.

The copper losses ($|I|^2 R$) in the primary
and secondary windings are practically independent of voltage. The controlling factor is
the current flow, and to keep these losses as
small as possible the coils are wound with
wire of the largest cross section that space
will permit. To minimize the total copper

losses, it is also necessary to proportion the wire size directly to the current flow in each coil. If the secondary voltage is only half the primary voltage, the secondary current will be about twice the primary current, and the secondary winding must have wire of twice the cross section used for the primary coil. However, since the secondary has only half as many turns in this case, the volume of copper used for the two coils is practically the same. This equal-volume relationship holds true for practically all transformers.

The iron losses or core losses in a transformer also have two components. The repeated magnetizing and demagnetizing of the core produces a loss because of repeated realignment of the magnetic domains. This loss, known as hysteresis loss, is directly proportional to frequency and proportional to about the 1.8 power of the flux density. The use of silicon steel alloy for the magnetic circuit minimizes hysteresis loss. The changing magnetic field also induces circulating currents or eddy currents in the core material, and this loss is proportional to the square of the frequency and also proportional to the square of the flux density. To minimize eddy current loss, the core is constructed of laminations or layers of steel (usually of less than 0.5-mm thickness) that are clamped or bonded together into an apparently solid mass.

Transformer iron losses are independent of load current, but since the flux in the core is directly proportional to the voltage, they are affected by the voltage applied to the primary winding. However, most power transformers operate at constant voltage and frequency, and it is then permissible to regard the iron losses as being constant.

The energy removed from the electrical circuit by the copper and iron losses of the transformer is converted to heat in the coils and the core, respectively. Considerable heat can be stored in the copper and iron, but under given operating conditions the average temperature of the assembly will ultimately rise until the heat can be dissipated as rapidly as it is being produced by the losses; it is therefore normal for a transformer to operate at a temperature somewhat higher than the surrounding air. However, the maximum permissible temperature that the transformer may be allowed to reach is dictated by the temperature rating of the insulation used for the coils, and therefore the losses in the transformer must not be allowed to remain at excessively high values for too long a time period. The result is that the copper losses set a maximum permissible continuous current value, and the iron losses determine a maximum[1] voltage value. These two limitations are substantially independent of each other and independent of load power factor. This is the reason why transformers are rated in kilovolt-amperes (kVA) rather than kilowatts (kw).

Two things should be noted. First, since the limiting factor is heat, short-time overloads do no harm.[2] Second, any condition or alteration that helps to remove heat from the transformer (such as a lower ambient temperature) will increase the permissible load.

It is fairly easy to carry the heat away from a small transformer. Often radiation from the surface of the core and coils plus the effect of convection currents is all that we

---

[1]Because of iron saturation, operating a transformer above rated voltage tends to drastically increase the no-load current. This also sets a maximum permissible operating voltage.

[2]The ASA loading guides for transformers published by the American Standards Association are among the most authoritative sources of information on the subject of short-time loading.

require. However, for a given geometric shape the losses are proportional to the cube of one linear dimension, but surface area is proportional to only the square; so the larger the unit, the more difficult the heat-dissipation problem becomes. Transformers up to about 1500 kVA can be "dry types." The core and coils are simply enclosed in a sheet-metal cabinet, and ventilation openings are provided at the top and bottom to facilitate convection currents. In some cases thermostatically controlled fans are provided to force the circulation of air, and for a given core and coil assembly, fans will increase the permissible load by about 33%.

Small oil (or noninflammable liquid) filled transformers have the core and coils mounted near the bottom of a cylindrical tank and the latter filled with liquid to about three-quarters of its height. The liquid serves as an important part of the transformer insulation, but convection currents in the oil carry heat to the surface of the tank from whence it is radiated and/or carried away by convection currents in the air.

As transformer size increases, manufacturers are forced to corrugate the tank walls or possibly provide cooling fins or tubes. On large units a fan may be required to circulate air over the cooling fins or tubes, and radiators (larger than but similar to an automotive radiator) may be required. On very large units the internal liquid may be forced to circulate by a pump, and sometimes the oil is cooled by circulating cold water through a series of tubes immersed in the oil.

The term efficiency (η) means the ratio of (output power in desired form) ÷ (input power). The heat produced in the core and coils of a transformer is seldom of any commercial value (which is why we refer to copper and iron losses instead of output), so

we normally only consider the secondary electrical output in watts and the primary input expressed in watts.

Mathematically,

$$\eta = \frac{\text{secondary watts}}{\text{primary watts}} \qquad (3\text{--}13)$$

Since, by definition, loss is the difference between the input and output watts, we can write Equation 3–14 and transpose it to Equation 3–15.

$$\text{losses} = \text{input} - \text{output} \qquad (3\text{--}14)$$

$$\text{input} - \text{losses} = \text{output} \qquad (3\text{--}15)$$

Using Equations 3–13 and 3–15, we can derive Equation 3–16.

$$\eta = 1 - \frac{\text{losses}}{\text{input}} \qquad (3\text{--}16)$$

Equations 3–13 and 3–16 give efficiency as a decimal (i.e., in per unit terms). To get percentage, multiply by 100.

The efficiency of most transformers is very high. Typical values are from 97% for a 1-kVA unit to over 99% for 1000-kVA or larger sizes, at unity power factor and full load. At a constant volt–ampere load, the percentage loss of a transformer is inversely proportional to the load power factor, so that if a given unit is 97% efficient at unity power factor (i.e., 3% loss) it will be only 94% efficient at 50% power factor (6% loss). Efficiency is also seriously affected by either over- or underloading. At high loads the efficiency falls off because the copper losses increase as the square of the load. At light loads the efficiency is poor because of the constant iron loss. The best efficiency occurs when the copper losses equal the iron losses, and this generally occurs somewhere between one-half and three-quarters of full

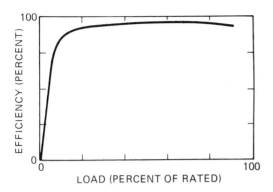

FIGURE 3-6 Graph of typical transformer efficiency

load, but the efficiency remains quite high down to about 10% of full load. A typical graph of transformer efficiency is shown in Figure 3-6.

The majority of transformers on power systems do not run at a constant load, but instead the load varies over a regular pattern every day. Under these conditions the full-load efficiency is not as important as the overall efficiency during a 24-hour day. The exact value of this 24-hour or all-day efficiency depends on the pattern of the load variation and methods of calculating it are treated later, but it cannot be greater than the maximum efficiency and usually is considerably lower.

Insulation Classes and Temperature Rise

The insulation in a transformer performs the threefold duty of insulating each turn of each coil from every other turn, insulating the coils from each other, and insulating each coil from the core (or from ground). When manufactured, the wire used to wind the coils is usually covered with some kind of insulating material, either cotton or a hard enamel-like material that goes by various trade names, like Formex or Formel.

The voltage between turns is comparatively low, and although the insulation on the conductors is very thin, it is adequate for the purpose. The voltage between coils and/or the voltage between each coil and the core is normally much higher than the turn-to-turn voltage. For this reason, additional solid forms of insulation are usually found between the coils and also between the coils and the core. In addition, various insulating spacers may be required to provide one or more places where the cooling medium can circulate between the coils. It is easy to see the insulating requirements, but we tend to overlook the question of mechanical support for the coils. The individual turns in any one coil tend to be crushed together when current flows through them, and there are rather large repulsion forces between the primary and secondary coils. These forces are not constant but are proportional to the square of the instantaneous current, and hence they are vibratory in nature under normal conditions and extremely high under short-circuit conditions. These vibratory forces are one reason why the coils (and sometimes the entire core and coil assembly) are dipped in an insulating varnish and baked into a solid mass. The idea is to glue every wire to its neighbors so as to prevent any vibration that could wear the insulation off two adjacent conductors and create a short circuit. With regard to the forces generated between coils, insulation spacers, wedges, and the like are required to prevent movements, because if the coils move, they will wear out the insulation at some point and a short circuit or a ground fault will result.

The material used for insulation may be

TABLE 3-1   INSULATION CLASSES AND TEMPERATURE RATINGS

Insulation Class	A	B	F	H
		°C		
Ambient	40	40	40	40
Rise by resistance	60	80	105	125
Hot spot allowance	5	10	10	15
Maximum hottest spot temperature	105	130	155	180

organic material such as paper, cotton, fiber, Micarta, or varnished cambric, or it may be inorganic material such as mica or glass fibers held together by some kind of bonding resin. But no matter what kind of insulation system is used, the mechanical properties of the insulation deteriorate with time. The deterioration process is basically a chemical action occurring within the material, and iike all chemical actions it is accelerated by heat and approximately doubles its rate for each 8° to 10°C increase in temperature. When the insulation deteriorates, the transformer develops loose coils and/or loose conductors within a coil, mechanical wear begins, and short circuits or ground faults soon result.

The National Electrical Manufacturers Association has grouped various types of insulation into classes and assigned a maximum permissible hottest spot temperature to each class. Since the hottest spot temperature is usually at some inaccessible spot within a coil, the maximum permissible average temperature (found by measuring the resistance of the coil) is somewhat lower. The difference between the ambient temperature and the average temperature of the coil is its temperature rise. The sum of the ambient temperature plus the maximum temperature rise plus the hot-spot allowance equals the maximum temperature rating of the insula-

tion. Table 3-1 shows the maximum hottest spot temperature and the maximum permissible temperature rise (based upon a 40°C ambient) for each class.

Mutual and Leakage Flux

At no load, almost all the flux set up by the primary follows the iron core and links all the turns of the secondary. But when the secondary is carrying current, the opposing mmf of the secondary has the effect of lowering the permeance of the magnetic circuit, and the result is that some of the flux created by the primary will take a path through the air instead of linking the secondary coil, as shown in Figure 3-7. The flux that links both windings is known as mutual flux, and it creates the secondary voltage as well as the primary cemf. But the flux that links only one winding is known as leakage flux, and its effect is the same as having some external inductive reactance in series with the primary coil. Transformer reactance (or inductive reactance or leakage reactance) is simply the opposition to current flow caused by the leakage flux within the transformer. In small units this reactance may be smaller than the resistance of the winding, but in large transformers it may be ten or more times the winding resistance.

Impedance Voltage, Short-Circuit Current, and Voltage Regulation

Although resistance and reactance exist in both windings, it is convenient to consider the equivalent impedance as being lumped together on one side or the other. This equivalent impedance is important because it determines the short-circuit current available

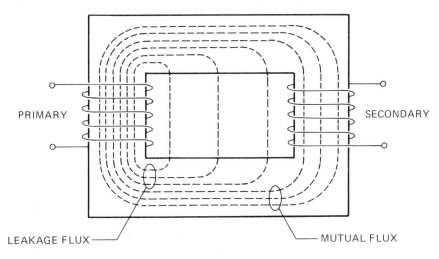

FIGURE 3-7 Mutual and leakage flux in a transformer

from and the voltage regulation characteristics of the transformer.

Impedance voltage ($|IZ|$) is the voltage required to circulate full-load current through the equivalent impedance of the transformer, and it is usually given as a percentage of rated voltage ($|IZ\%|$). When given as a percentage, impedance voltage is the same on either the primary or secondary sides. If desired, impedance voltage can be measured by short circuiting one side of the transformer and applying sufficient voltage on the other side to circulate full-load current. The voltage required to do this divided by the rated voltage on the energized side gives $|IZ|$ in per unit terms (multiply by 100 to get percent).

If rated voltage is maintained on the primary and the secondary in short circuited, the current that flows can be easily calculated if $|IZ\%|$ is known. Equation 3–17 shows how. Note the use of absolute rms values for currents (i.e., ammeter readings), not complex notation.

$$|I_{sc}| = |I_{fl}| \times \frac{100}{|IZ\%|} \qquad (3\text{--}17)$$

where $|I_{sc}|$ = short-circuit current on the secondary or primary side

$|I_{fl}|$ = full-load current on the secondary or primary side

Voltage regulation is the change of voltage caused by a change from full-load to no-load conditions with all other factors remaining constant. It may be quoted in volts but is usually expressed as a percentage of the secondary voltage at full load. Mathematically, using absolute rms voltage (voltmeter measurements), and not complex notation,

$$V \times \text{reg. }\% = \frac{|V_{nl}| - |V_{fl}|}{|V_{fl}|} \times 100 \qquad (3\text{--}18)$$

where $|V_{nl}|$ = no-load secondary voltage

$|V_{fl}|$ = full-load secondary voltage

When load is applied to or removed from

a transformer, the voltage drop due to the internal impedance ($|IZ|$) causes the secondary voltage to change. So one might expect the voltage regulation (%) to equal $|IZ\%|$, but such is the case only if the load power factor is at the most unfavorable (lagging) value. At any other load power factor, voltage regulation will be less than $|IZ|$. Calculations are dealt with later, but it is useful to know that, with a leading power factor load, voltage regulation can become zero or even negative. To state the limits mathematically,

$$|IZ\%| > V \text{ reg. } \% > -|IZ\%| \quad (3\text{-}19)$$

As defined here, voltage regulation is a circuit characteristic, and this is the most common usage of the term. Sometimes "voltage regulation" is used to indicate the "action of controlling the voltage at some point in a circuit," but that usage does not appear in this book.

Standard Terminal Markings and Polarity of Single-Phase Transformers

North American manufacturers have adopted the following standard terminal markings for transformers. The high-voltage leads or terminals are identified as H_1 and H_2, and the low-voltage leads or terminals are identified as X_1 and X_2. The terminals are marked so that (considering instantaneous voltages across the winding) when H_1 is positive in relation to H_2, then X_1 will be positive in relation to X_2.

Whenever practicable, transformers are constructed with the high-voltage terminals mounted on one side and the low-voltage terminals on the other side. In these cases, viewing the transformer from the high-

voltage side, it is standard practice to mark the right-hand high voltage terminal H_1.

The terminal markings on transformers may be stamped or painted next to the terminals, or the nameplate may include a sketch on which the terminals are identified. Flexible leads must be identified with tags or tape on the insulation, or numbers stamped into the lugs on the ends of the leads.

It can be shown that, if two terminals (such as H_1 and X_1) on two different windings of a transformer inherently become positive at the same time, then current flow into the transformer on either terminal H_1 or X_1 will magnetize the core in the same direction; if one of the windings is primary and the other secondary, then (neglecting magnetizing current) the instantaneous current flow will always be into the transformer on one terminal (H_1) and out on the other (X_1).

The term "polarity" refers to the positions in which the transformer secondary terminals are mounted relative to the primary, and cannot be applied to a unit that has flexible leads instead of fixed terminals. Specifically, if terminal X_1 is mounted closest to H_1, the transformer has subtractive polarity, and if X_2 is closest to H_1, it has additive polarity. The terms "additive" and "subtractive" polarity originate from the polarity test used by linemen in the field before terminal markings were standardized. The test is performed by joining one high-voltage terminal to the *nearest* low-voltage terminal with a wire jumper and connecting a voltmeter between the remaining H and X terminals, as shown in Figure 3-8(a), (b), (c), or (d). Energize the high-voltage winding with a convenient ac voltage and observe the voltmeter. If the voltmeter indication is less than the applied voltage, the transformer has sub-

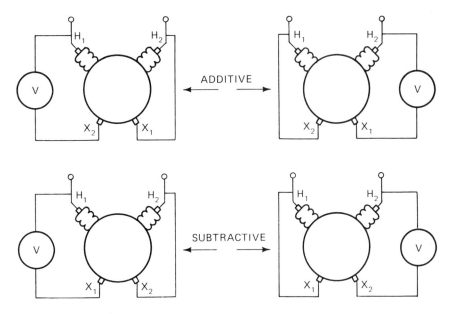

FIGURE 3-8   Polarity tests on a distribution transformer

tractive polarity, as shown in Figure 3-8(c) and (d). If the voltmeter indication is more than the applied voltage, the transformer has additive polarity, as shown in Figure 3-8(a) and (b). Notice that for this test it does not matter whether the jumper is on H_1 or H_2 as long as it is connected to the *nearest* low-voltage terminal. It is always possible to obtain an additive voltmeter indication on a transformer with subtractive polarity (or vice versa) by interchanging the two wires connected to the low-voltage winding, but this may lead to confusion and *does not change the transformer polarity*.

Canadian manufacturers have standardized on additive polarity for distribution transformers up to 500-kVA size and up to 34.5-kV rating (see CSA standard C2-1969, Single-phase and Three-phase Distribution Transformers, Types ONAN and LNAN).

Parallel Operation of Transformers

For purposes of the following discussion, let us define parallel operation as an arrangement where the primaries of two or more transformers are energized from the same source (i.e., the same generator or the same output winding of a bigger transformer or voltage regulator), and the secondaries are connected to supply the same load. Figure 3-9 shows a schematic diagram of two transformers in parallel. There may be considerable lengths of independent circuit wiring (i.e., separate conductors or cables) supplying the primaries, and the secondary wiring may be similarly arranged, but separate overcurrent protection is usually applied to each unit, often on both the primary and secondary sides.

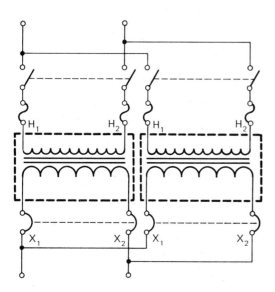

FIGURE 3-9 Two transformers connected in parallel

The effect of line impedance is neglected in this discussion, and any possibility that the transformers might serve as a tie between isolated sections of the primary system, or that as a part of normal operation, power flow into one transformer primary, across between secondaries, and out of another transformer primary might be desirable, is specifically excluded.

Parallel operation of transformers is one practical way to increase the capacity of an existing installation and also improve system reliability. But it will increase the available short-circuit current, sometimes to a level beyond the interrupting rating of the existing overcurrent devices. A split secondary bus with a tie breaker interlocked to prevent parallel operation may be necessary in such a case.

In addition to the usual consideration of voltage rating, kVA capacity and available short-circuit current, the following recommendations should be adhered to when paralleling transformers.

Matched Secondary Voltages

The no-load secondary voltages should be exactly equal. If they are not, the difference will cause current to circulate between the secondary windings (and also between the primaries). This causes unnecessary transformer heating (sometimes quite noticeable at no load), and therefore will reduce the maximum permissible load to something less than the sum of the transformer nameplate ratings.

Different makes or models of transformers may have slightly different secondary voltages even though their nameplate voltage ratings are the same, so this point should be carefully checked. If the transformers have taps, it is obvious that the same tap setting must be used on both units.

Matched Impedances

The transformers should have the same impedance drop in percent so that they will share the load in proportion to their kVA ratings. If the transformer impedances are not equal, the one with the smallest $IZ\%$ will carry more than its share of the load. As the total load increases this transformer will reach full load, while the other transformer with more impedance will still be operating at part load. So the maximum permissible load will be less than the sum of the kVA ratings of the units.

Matched Impedance Angles

The X/R ratio for both transformers should be the same. Otherwise their secondary currents will be somewhat out of phase with each other, and the maximum permissible load will be less than the sum of the nameplate ratings.

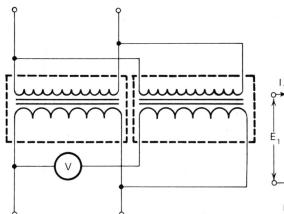

FIGURE 3-10 Phasing out transformers

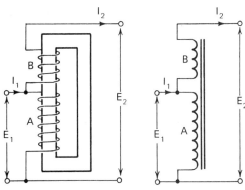

FIGURE 3-11 Sketch and schematic of a boosting autotransformer

Correctly Phased Out

The secondary windings must be connected so that (1) the output voltages will be in phase opposition around the loop between transformers *or*, to express this another way, (2) the output voltages will be in phase with each other viewed from the load. Notice in the preceding statements that it does not matter how the primary windings are connected as long as the secondaries are properly phased out. If phasing out the secondaries cannot be accomplished visually, then make a trial connection with a voltmeter as shown in Figure 3-10, and energize both transformers. If the voltmeter indicates zero (or very nearly so), the connection is correct. If the voltmeter indicates twice the secondary voltage of one transformer, the incoming unit must be reversed (interchange either the primary *or* the secondary line wires, but not both).

Once it is known how the secondaries should be connected, make the permanent connections and check for circulating current at no load and proper division of current at or near full load.

Note that the first three requirements for parallel operation must be built into the transformer by the manufacturer. Transformers of the same make and model and the same kVA rating will usually parallel satisfactorily, but when ordering a transformer to be paralleled with an existing unit, advise the manufacturer of this intent. It *is* possible to force satisfactory load division between dissimilar transformers in parallel, but we will not discuss that problem here.

Whenever transformers are connected into parallel banks, disconnecting means must be provided on both the primary and secondary sides of each transformer, and both the associated disconnect switches must be open before any work is done on any unit. If only the primary switch is opened, a feedback occurs on the secondary side, and rated voltage will still exist on both the primary and secondary sides of that transformer.

Autotransformers

Although other configurations are possible, the following discussion will be confined to autotransformers like those shown in Figures 3-11 and 3-12. Although it is often defined

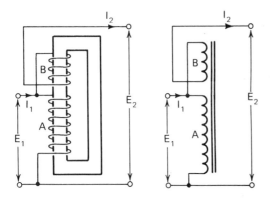

FIGURE 3-12   Sketch and schematic of a bucking
autotransformer

as having only one winding, an autotransformer often physically resembles two distinct coils connected together, and we shall analyze its operation from a two-coil viewpoint. For convenient reference we will identify the coils as the common winding (A) and the extension winding (B) in Figures 3-11 and 3-12.

Consider the autotransformer sketch in Figure 3-11. If the voltage $|V_1|$ is applied to the common winding as shown, the voltage induced in the extension winding will add to $|V_1|$ to produce the output voltage $|V_2|$. As long as the right-hand terminals are open circuited, only a small magnetizing current will flow in coil A, just enough to produce the appropriate cemf (practically equal and opposite to V_1) in that coil. The voltage boost provided by coil B is just $|(V_2| - |V_1)|$ and will be determined by the turns ratio between the common and extension coils. If a load is connected to the right-hand terminals, the current I_2 flowing through coil B exerts an mmf on the core; but because the winding resistances are small, the cemf cannot change much. Therefore, the flux in the core cannot be altered, and so the total ampere turns on the core must remain almost constant. The current in coil A therefore must be in-

versely proportional to the turns ratio of the coils and flow in such a direction as to offset the mmf of coil B (i.e., opposite wrap directions around the core). Neglecting magnetizing current, the same argument about constant total ampere turns on the core holds true for the bucking autotransformer shown in Figure 3-12.

$$|I_{\text{coil } A}| = |I_2| \frac{N_B}{N_A} \qquad (3\text{-}20)$$

$$|I_1| = |I_2| + |I_{\text{coil } A}| \qquad (3\text{-}21)$$

However, neglecting magnetizing currents,

$$|V_2| = |V_1| - |V_{\text{coil } B}| \qquad (3\text{-}22)$$

and $\qquad |I_1| = |I_2| - |I_{\text{coil } A}| \qquad (3\text{-}23)$

Autotransformers cost less, have better voltage regulation, and are smaller, lighter in weight, and more efficient than mutual-induction transformers of the same kVA capacity. These advantages become greater as the ratio between the primary and secondary line voltages $|V_1/V_2|$ approaches unity.

Compared to a mutual-inductance type, an autotransformer has two disadvantages. The latter does not isolate the secondary from the primary circuit. This may make it unsuitable for some purposes, but it also means that, if the common winding becomes open circuited, the primary voltage can still feed through the extension winding to the load. With a step-down autotransformer this could result in burned out secondary loads and/or a serious shock hazard, particularly if the step-down ratio were high.

Autotransformers are often used on power transmission circuits, but their use on indoor wiring systems is limited. Common applications are fluorescent lamp ballasts, reduced-voltage motor starters, and series line boosters for fixed adjustment of line voltage.

Connections for Three-Phase Transformer Banks

Although other possibilities exist, the following discussion refers primarily to single-phase two-winding transformers that are connected into groups or banks to transform three-phase voltage. Specifically excluded are three-phase transformers (manufactured as such) that have three-legged cores, and any transformer or bank that has more than one input or more than one output three-phase circuit. The viewpoint is one of using some existing transformers for a given application.

There are only four practical ways of connecting three single-phase two-winding transformers to make up a three-phase bank (i.e., wye–wye, wye–delta, delta–wye, and delta–delta). The connection required for a given installation depends upon the voltage rating of the transformer to be used, the primary voltage available, and the secondary

voltage desired. These considerations usually narrow the range of possible connections down to one of the four, but if any choices remain, use whatever connection will result in the highest practicable voltage across each transformer, as this will result in maximum kVA capacity for the bank. The open delta–open delta and the open wye–open delta connections permit transformation of three-phase power using only two transformers. Each of these possible connections has some unique characteristics and advantages, some of which are presented here. But a complete discussion of the pro's and con's of each is beyond the scope of this book.

Wye–Wye Connection

It is standard practice to connect wye–wye banks so that there will be no phase displacement between the primary and secondary line-to-line voltages, as shown in Figure 3-13, but it is possible to shift the secondary voltages 180° by reversing all three secondary

FIGURE 3-13 Two schematics and a sketch of a wye-wye transformer bank without a primary neutral

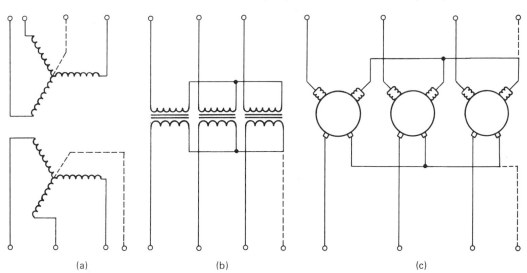

(a) (b) (c)

coils. The neutral connections are optional, but their inclusion or omission gives the bank different characteristics.

WITHOUT A PRIMARY NEUTRAL   If transformers that normally require different amounts of magnetizing current are wye connected with no primary neutral, the voltages across the transformers will be unbalanced under no-load conditions. The transformer requiring the least magnetizing current will have more than normal voltage across it, possibly enough to cause overheating. The transformers used should therefore be identical.

Even if the secondary line-to-wye point voltages (measured at no load) appear suitable, it is not practical to use the wye point as a neutral connection for a secondary four-wire wye system. If a single-phase load is connected from one line to the wye point on the secondary side, the voltage across the loaded transformer decreases appreciably, and a comparable voltage increase occurs across the other units, which may cause them to overheat. These voltage changes are observable on both the primary and secondary sides of the transformers and so are not to be confused with or attributed to the normal voltage-regulation characteristics of the units. The severe voltage unbalance that occurs with unbalanced line-to-wye point secondary loads is known as neutral instability and makes this kind of loading impractical, even though the line-to-line voltages remain normal. Single-phase line-to-line loads do not have the same effect on the voltages.

The transformers may be subject to harmonic overvoltages if the supply system is isolated and the primary wye point is grounded.

WITH A PRIMARY NEUTRAL   The addition of a primary neutral connection makes each transformer independent of the other two. Dissimilar transformers will not cause voltage unbalance at no load and therefore may be used in this case.

Secondary neutral instability is also eliminated, so the bank can be used to supply a secondary four-wire wye wiring system. However, the harmonic components of magnetizing current flowing in the primary lines can cause interference on nearby telephone circuits, particularly if the earth is used as a neutral conductor.

Wye–Delta Connection

Connections for a typical wye–delta transformer bank are shown in Figure 3-14, but this is not the only possible connection. The secondary line-to-line voltages are always shifted 30° from the primary line-to-line voltages, but this is important only if parallel operation is to be attempted with some other transformer bank. Here again the neutral connection is optional, but its presence or absence gives the transformer bank different characteristics.

WITHOUT A PRIMARY NEUTRAL   Transformers connected in this fashion need not be identical. Differences of magnetizing current requirements are offset by a small current circulating around the delta, and no appreciable voltage unbalance can occur.

The isolated primary wye connection also forces the transformers to uniformly share a balanced load, even if the transformers do not have equal impedance. The secondary may be used to supply a four-wire delta wiring system if a center tap is available on one transformer.

If one primary line becomes open (e.g., owing to a blown primary fuse), three-phase

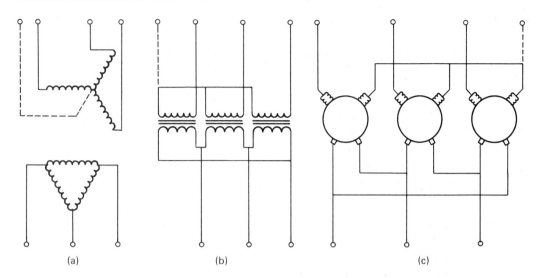

(a)   (b)   (c)

FIGURE 3-14   Two schematics and a sketch of a wye-delta transformer bank

motors that are operating on the secondary side of the bank will continue to run, drawing normal current on two of the lines and about twice normal current on the third wire. It requires three overload devices to properly protect a motor against this condition.

WITH A PRIMARY NEUTRAL   The transformer bank can supply either three-phase three-wire or four-wire delta circuits; in fact, the bank will still deliver three-phase power with one transformer entirely disconnected from the other two, although their capacity will be only 57.7% of the original capacity of the bank.

However, transformers connected wye-delta with a primary neutral tend to share the load inversely to their internal impedances so that identical transformers should be used. Even with identical units, unbalanced primary line-to-neutral voltages can still cause uneven sharing of load current with consequent overheating of one or two of the transformers.

Any wye–delta transformer bank will introduce a 30° displacement between the primary and secondary line-to-line voltages. The secondary voltages may either lead or lag the primary.

Delta-Wye Connection

A delta–wye transformer bank can supply a secondary four-wire three-phase wye system. There is no problem regarding magnetizing currents, and the secondary wye connection forces equal sharing of balanced secondary loads, so the transformers need not be identical.

The delta–wye connection also introduces a 30° phase shift (either leading or lagging) between the primary and secondary line-to-line voltages.

Typical connections are shown in Figure 3-15.

Delta-Delta Connection

Typical connections for a delta–delta transformer bank are shown in Figure 3-16. If con-

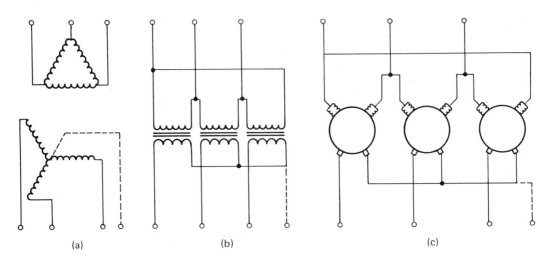

FIGURE 3-15   Two schematics and a sketch of a delta-wye transformer bank

nected as shown, the primary and secondary voltages will be in phase with each other, but reversing all the secondary coils will shift the secondary voltages 180°.

If a center tap is available on one transformer, the bank may be used to supply a four-wire delta wiring system.

If one transformer fails in service, the remaining two can be operated as an open delta at 57.7% of the original capacity of the bank. However, in a complete delta–delta bank, the transformers tend to share the load inversely to their internal impedances and so identical transformers should be used.

Open Delta–Open Delta Connection

As can be seen from Figure 3-17, this connection is basically a delta–delta from which one transformer has been removed. It is used

FIGURE 3-16   Two schematics and a sketch of a delta-delta transformer bank

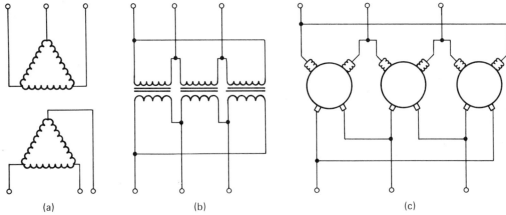

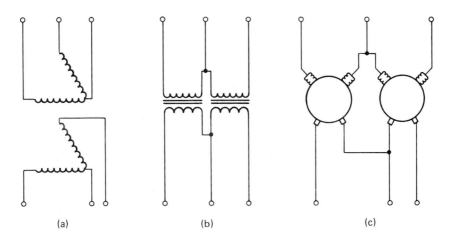

(a) (b) (c)

FIGURE 3-17 Two schematics and a sketch of an open delta–open delta transformer bank

to supply three-phase three-wire or four-wire delta secondary systems. Its main advantage is the reduced investment required for a small three-phase bank, and it is not economical for large installations. The secondary voltages tend to unbalance slightly under load.

If used to supply a four-wire delta secondary system, the two transformers used are often of different sizes. The single-phase load is always connected across the larger transformer. If the power factor of the single-phase load is higher than that of the three-phase load, the transformers should be connected to the primary supply so that the voltage across the large transformer leads the voltage across the other by 120°.

With equal-sized transformers operating at rated voltage and a balanced three-phase load, the capacity of the bank will be 86.6% of the sum of the transformer ratings.

Open Wye–Open Delta Connection

As can be seen in Figure 3-18, this connection is basically a wye–delta connection from which one transformer has been omitted. The primary neutral connection is essential. This connection does not balance the load across all three primary lines, but otherwise the remarks made about the open delta–open delta apply equally well here.

Three-Phase Transformer Assemblies

Because they offer greater economy of size, weight, and cost per kVA, three-phase transformers are often preferable to three single-phase units. The magnetic circuit of a three-phase unit is usually such that certain portions of it are common to two or more phases, but the coils are physically arranged in two or three single-phase sets.

The coils may be connected in wye–wye, wye–delta, delta–wye, delta–delta, or T-T, but open deltas or open wyes are rarely if ever used. The internal connections are frequently shown on the transformer nameplate.

The terminals of a three-phase transformer are usually marked in accordance with the following rules:

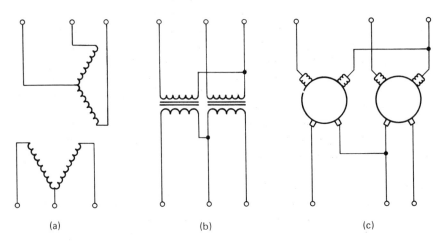

      (a)   (b)   (c)

FIGURE 3-18   Two schematics and a sketch of an open wye–open delta transformer bank

- The high voltage line terminals are marked H_1, H_2, and H_3. If a neutral connection is provided, it is marked H_0.
- The low-voltage line terminals are marked X_1, X_2, and X_3, with X_0 being the neutral connection if such is provided.
- If a neutral connection is provided, it normally will be at (or is intended to be at) a potential such that all three lines to neutral voltages are balanced (i.e., equivalent to a four-wire wye).
- If the high-side phase sequence is 1–2–3, the low-side phase sequence will be 1–2–3.
- If possible, the primary and secondary line voltages should be in phase, but, if not, the secondary voltages should lag by 30° (compared on a number-to-number basis).

Instrument Transformers

Because of their size and delicate construction, it is rather difficult to build ammeters or voltmeters that can be directly connected into circuits over 750 V, and it is also difficult to design ammeters or kilowatt-hour meters for high currents. Where high currents or high voltages are involved, these de-

vices are operated by small transformers known as instrument transformers that serve no other useful purpose.

Potential transformers for use up to 15 kV are not unlike the single-phase transformers that have already been described, but they are rather small, often of less than 50-VA capacity. Potential transformers for higher voltages are sometimes a capacitive voltage divider and transformer assembly, as shown in Figure 3-19, and these usually have a higher VA rating. But no matter what the exact construction may be, the unit must have a fixed and accurately known ratio between the input and output voltages and the

FIGURE 3-19   Potential transformer rated at 138kV/120 V

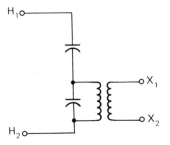

minimum practicable phase difference between them. Usually the output voltage is nominally 120 V no matter what primary voltage rating has been selected (e.g., ratings of 2400/120 V or 14,400/120 V are common), and most of them can be operated at up to 25% higher voltage (maximum of 150-V output).

Voltmeters used with potential transformers are usually designed for full-scale deflection with 150 V applied. The scale plate is simply calibrated so that the meter directly indicates the voltage on the primary side of the potential transformer.

A current transformer follows the same theory as any other transformer, but the construction is unique and so is the way it is used. A sketch of a current transformer as it is generally used in a circuit and a schematic diagram of the same arrangement are shown in Figure 3-20. Note that the current transformer is wired in series with the power circuit load. Because the primary winding has very few turns and the core is small, very little voltage drop occurs on the current transformer primary. The current flow is determined only by the load impedance.

But because of the turns ratio, the current to the ammeter will bear a fixed ratio to the primary current, and most current transformers are designed for a secondary current of 5 A. So we generally use a 5-A meter movement and mark the scale in such a way as to take the transformer ratio into account.

As long as the current transformer secondary circuit is closed, the secondary mmf opposes the primary mmf, the flux in the core is at a low value, and the secondary voltage is low (generally below 5 V). But if the secondary circuit is opened, the primary current does not change significantly, the flux level rises to saturation values, and the secondary voltage becomes very high, some-

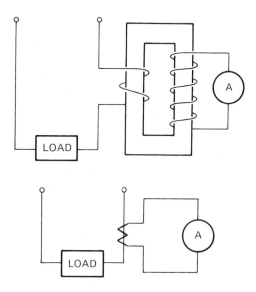

FIGURE 3-20 Sketch and schematic of a current transformer in a typical power circuit

times high enough to break down the insulation and nearly always high enough to create a serious shock hazard. It is common practice for electrical tradesmen to work on the secondary circuit of a live current transformer, but because of the danger of an accidental opening, the current transformer secondary should be short circuited right at its terminals before any work is begun. Most current transformers have some convenient arrangement for doing this. Never fail to use the short-circuiting device, and do not forget to remove it when the work has been completed. The current transformer shown in Figure 3-20 has a wound primary coil, but this is not true of all current transformers. "Doughnut" types of current transformers that have only a core and a wound secondary are quite common. The circuit conductor is routed (once only) through the window of the core, and this serves as a one-turn primary winding.

Both potential and current transformers

are subject to ratio errors (e.g., the ratio between the input and output voltages or currents may be slightly different from the nominal value) and phase-angle errors (e.g., a slight phase shift may exist between the input and output voltages or currents). Voltage and current measurements taken through instrument transformers are affected only by the ratio errors. But a kilowatt-hour meter is commonly operated by potential and/or current transformers, and it is affected by both ratio and phase-angle errors. The ratio errors add together, and if the transformers all have 0.25% ratio errors with the output being low, the kilowatt-hour meter reads 0.5% low. With the phase-angle errors the angles still add, but the net effect depends upon the power factor of the load that is being measured. At unity power factor, phase-angle errors have almost no effect on a wattmeter or a kilowatt-hour meter; but if the load power factor is very low, the effect may be substantial. If the phase angle of the metered load is 85° for example, a phase-angle error of 0.5° in a current or potential transformer will create about a 10% error in the wattmeter reading obtained.

The ratio and phase-angle errors of a potential transformer are mainly influenced by the magnitude of its secondary load, commonly known as its burden. The same is true of current transformers, but their accuracy is also affected by the load current.

The accuracy of an instrument transformer is specified in terms of the error it may produce on a wattmeter that is measuring a 50% power factor lagging load. It is assumed that the total burden on the in-

TABLE 3-2   STANDARD BURDENS FOR INSTRUMENT TRANSFORMERS

	Standard Burden Characteristics		CURRENT TRANSFORMERS	For 60-Cycle and 5-Ampere Secondary Current	
Burden Designation	Resistance: Ohms	Inductance: Millihenrys	Impedance: Ohms	Volt-Amperes	Power Factor
B-0.1	0.09	0.116	0.1	2.5	0.9
B-0.2	0.18	0.232	0.2	5.0	0.9
B-0.5	0.45	0.580	0.5	12.5	0.9
B-1	0.5	2.3	1.0	25	0.5
B-2	1.0	4.6	2.0	50	0.5
B-4	2.0	9.2	4.0	100	0.5
B-8	4.0	18.4	8.0	200	0.5

	POTENTIAL TRANSFORMERS				
Burden Designation	Secondary Volt-Amperes	Burden Power Factor			
W	12.5	0.10			
X	25	0.70			
Y	75	0.85			
Z	200	0.85			
ZZ	400	0.85			

strument transformer is some specified value, and in the case of a current transformer, the current is assumed to be at full rated value. The accuracy classes are 0.3, 0.6, or 1.2, and these numbers indicate the maximum percent error. If the power factor of the metered load is higher, or if the burden on the instrument transformer is lower, the accuracy tends to improve. In the case of the current trans-

former, the percentage error at 10% of rated current may be twice that allowed at full load.

Table 3-2 shows some standard burdens with which the accuracy of an instrument transformer may be specified. Typical statements of accuracy are "Accuracy class 0.6 B-0.5" for a current transformer or "Accuracy class 0.3w" for a potential transformer.

SOME ASPECTS OF TRANSFORMER DESIGN

Transformers used by the electric power industry are not quite as simple as the sketches that have so far appeared in this book would indicate. There are quite a few constructional refinements intended to improve efficiency, decrease voltage regulation, provide alternative voltages or voltage adjustment, improve cooling, decrease manufacturing costs, or to cope with mechanical forces. We will examine some of these refinements in the following pages.

Coil Configurations

Although Figure 3-1 shows the primary and secondary coils physically separated (wound on different legs of the core), this is rarely if ever the case. The problem is that such physical separation permits a great deal of leakage flux, and the voltage regulation becomes excessive. The usual arrangement is to put half of the secondary winding on each leg and then place half of the primary winding around the outside of each secondary coil. There normally will be a considerable thickness of insulation between the primary and secondary coils, and there usually are spacers between the coils to permit more effective cooling; nevertheless, this concentric coil

design reduces the area through which leakage flux can travel, so the voltage regulation is improved.

In practically all transformers the coils are wound on a form, dipped in insulating varnish, baked into a rigid mass, and then are assembled with the core. It is easiest to build the core with legs of rectangular cross section, and so it is most economical to wind the coils on a rectangular form. On small transformers rectangular concentric coils are practical, but in a large unit, the high repulsion forces generated between the primary and secondary coils under short-circuit conditions tend to "round out" the flat sides of the outer coil; if this happens, the resulting damage to the coil insulation will usually make the transformer unserviceable. Cylindrical concentric coils are the obvious answer to this problem of mechanical strength, but if they are used, the core legs usually have a more or less round cross section to best utilize the available space.

The simplest concentric coils are wound in layers. For small current values (10 A or so), round wire may be used, but wire of rectangular or square cross section will make better use of the available space and is preferred for larger sizes. Sometimes it is most economical to wind each primary and each secondary coil as one continuous layer-wound coil, but sometimes it is preferable to wind each coil as a series of narrow disc-shaped coils that are then stacked into a column and connected in series. This is known as a wire-wound disc-type coil. The wire is usually of round cross section, and this construction is normally found in small high-voltage transformers.

If a transformer coil is to be wound using wire of fairly large rectangular cross section, it becomes practical to wind the turns in the form of radial discs, rather than axial layers. The discs may be spaced from each other for improved cooling, but otherwise the continuous wound disc coil is similar to a layer type.

The design generally indicated by the term "pancake coil" is different. Here, the primary and secondary coils are each composed of a series of disc-shaped coils that are only the width of a single conductor. These discs are stacked alternately into a column and assembled with the core, and the interconnections between the coils to form complete windings are made later. The completed windings are described as interleaved pancake windings.

Magnetic Circuit Construction for Single-Phase Transformers

The simplest magnetic circuit used for single-phase transformers is that shown in Figure 3-1, and those that use it are known as core-type transformers. The easiest way to build such a magnetic circuit is to stack L-shaped laminations together, as shown in Figure 3-21, but other arrangements such as C-shaped and I-shaped laminations are also practical. In this core, as in most others, the individual pieces of iron are positioned

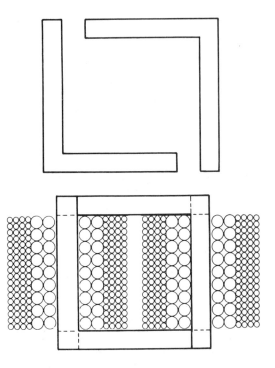

FIGURE 3-21   Core-type transformer: (top) individual laminations; (bottom) cross section of a core and coil assembly

so that the pieces in any one layer butt together, but joints in adjacent layers do not coincide. This improves the permeance of the core, and when the whole stack of laminations is clamped together, the core becomes quite rigid.

A shell-type transformer is shown in Figure 3-22. The figure shows E- and I-shaped laminations, but other designs, such as two E shapes, are possible. The diagram shows the high-voltage winding as being entirely wound overtop of the low winding, but it is common to find the low-voltage winding in two sections, with the high-voltage winding sandwiched between.

A distributed core is similar to the shell type but has four outer legs on the core. Typi-

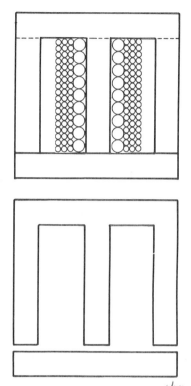

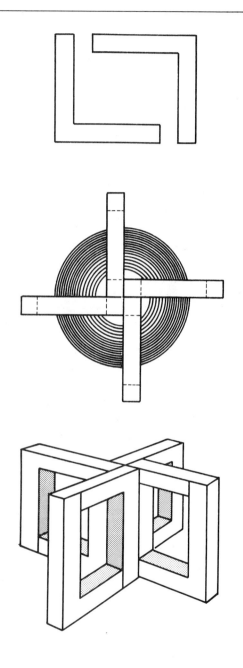

FIGURE 3-22 Shell-type transformer: (top) typical
laminations; (bottom) cross section of a core
and coil assembly

cal lamination shapes and the overall arrangement of the core and coils are shown in Figure 3-23. The advantage of this arrangement
is the reduced length of the magnetic circuit,
which reduces iron losses and no-load current.

If the laminations are all the same, it is obvious that the magnetic circuit will have a rectangular cross section. However, if slimmer
laminations are used for the outer layers, it is
possible to get a more or less round cross section in those legs on which the coils are
mounted, making these cores suitable for
cylindrical coils.

In an effort to decrease the iron losses and
magnetizing current of transformers, a great

FIGURE 3-23 Distributed-core tranformer: (top)
typical laminations; (center) top view of core and
coil; (bottom) core only

deal of research and development of magnetic core materials has been done over the years, and one of the more important developments is grain-oriented silicon steel. This material has a very high permeability in the rolling direction (i.e., in the direction in which it was rolled during manufacture), but not as good in any other direction. When used for transformer cores, punchings made from grain-oriented steel must be confined to I shapes (E's, C's, or L's are not suitable) cut in such a way that they will carry flux in the preferred magnetizing direction. Right-angle butt joints are not as good as 45° bevel cuts on the ends of the laminations or strips and are usually avoided.

Grain-oriented steel is inherently suitable for wound-core construction. The simplest way to build a wound core is to wind a continuous strip of steel on a rectangular form, bonding the turns together every inch of the way. When the desired thickness of core has been built up, split the core for convenient assembly with the coils and clamp the two halves together.

Figure 3-24 is a cross-sectional sketch of a wound-core transformer. The core shown is wound in two loops, which are split, assembled with the coils, and clamped together.

Magnetic Circuits for Three-Phase Transformers

The simplest arrangement is the three-legged core on which the coils are mounted side by side, as shown in Figure 3-25. It may be constructed of stacked laminations as shown, or it may be a wound-core design; in either case, the coils are mounted and connected symmetrically.

A three-phase transformer with a five-legged core is shown in Figure 3-26. It also may be either stacked laminations or a wound design, but the coils are always mounted and connected symmetrically.

Dual-Voltage Transformer Windings

To reduce the number of transformer designs (differing only in voltage rating) that must be produced, the manufacturers may divide the high- and/or the low-voltage windings into two identical sections or coils, as shown in Figure 3-27(a), (b), and (c).

The possibility of connecting the coils in either series or parallel enables the user to

FIGURE 3-24   Wound-core transformer

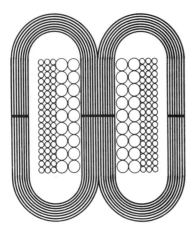

FIGURE 3-25   Three-phase transformer with a three-legged core

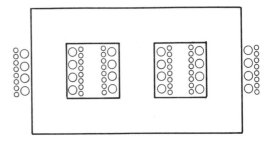

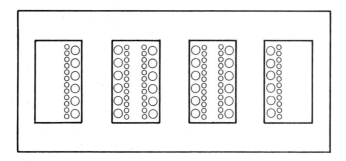

FIGURE 3-26 Vertical design of a three-phase transformer

obtain either of two optional voltages from a split secondary winding, and the winding may also be suitable for a three-wire single-phase output. Similarly, a split primary winding permits the use of either of two optional supply voltages. So a transformer designed as shown in Figure 3-27(c) can replace six single-voltage designs on which only the essential leads exist. However, to make the transformer suitable for unbalanced three-wire loads, it is essential that the secondary coils be closely interleaved with each other, rather than being located on different parts of the magnetic circuit. If the secondary coils are physically separated, they will work properly in series or in parallel, but when used for three-wire service, unbalanced loading will cause unbalanced voltages to appear at their terminals.

Because of the additional interleaving of the coils that is required, transformers designed for three-wire service are slightly more expensive than those intended for either series or parallel connection only, and so both types are found in industry.

If a transformer is not designed for three-wire service, the secondary voltages are quoted in the form (parallel $\times$ series) volts (e.g., 240 $\times$ 480 V). If the transformer has two separate secondary coils and is suitable for three-wire service, the voltages are quoted as (parallel–series) volts (e.g., 120–240 V). If the transformer has a center-tapped secondary as shown in Figure 3-27(a), the secondary voltage is specified as (whole–half) voltage (e.g., 240–120 V).

Figure 3-27 shows the standard numbering

FIGURE 3-27 Dual-voltage transformer windings

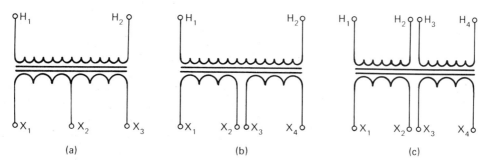

(a)   (b)   (c)

used for dual-voltage transformers. The phasing is such that the lower numbers are simultaneously positive on all coils, and this must be borne in mind when connecting the coils in either series or parallel.

Tapped Transformer Windings

Because the voltage applied to the primary of an installed transformer does not always exactly match the nameplate rating, many transformers have taps arranged in the high-voltage winding so that rated secondary voltage can be obtained when the primary voltage is moderately different from the nominal voltage rating of the unit.

Standard distribution transformers have taps arranged in $2\frac{1}{2}\%$ steps so that rated secondary voltage can be obtained when the primary supply is 0, $2\frac{1}{2}$, 5 or $7\frac{1}{2}\%$ below the nominal primary voltage rating. Transformers used in unit substations often have taps arranged in $2\frac{1}{2}\%$ steps and may include some overvoltage taps to accommodate primary voltages above the nominal rating. Taps in $1\frac{1}{4}\%$ steps may be used on transformers with automatic tap-changing equipment. Most transformers will carry rated kVA on any overvoltage tap and/or rated current on a reduced voltage tap.

Standard distribution transformers usually include a manually operated rotary switch for tap changing, as shown schematically in Figure 3-28. As a rule these switches are intended to be operated *only* with the transformer deenergized.

In unit substations, the taps are often brought out as different terminals to which the high-voltage lines can be connected, as shown schematically in Figure 3-29. Either of these designs is satisfactory as long as the supply voltage is reasonably constant. The

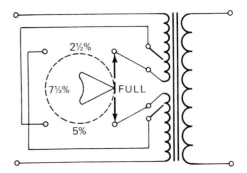

FIGURE 3-28   Distribution transformer with a no-load tap-changing switch

correct tap (or tap switch setting) can be chosen on the original installation, and a satisfactory secondary voltage thereby obtained. Either of these designs must be deenergized while changing taps, but on many installations, tap changes are very seldom required (once in several years). They are only needed as load growth and other system changes cumulatively alter the voltage at the load, so this type of no-load (or off-load) tap changing is adequate.

Transformers with On-Load Tap Changers

If load current variations cause an objectionable variation of system voltage, some method

FIGURE 3-29   Transformer with alternative line connections

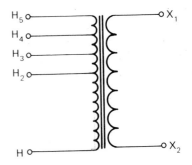

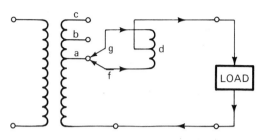

FIGURE 3-30   Elementary design of transformer for
on-load tap changing

of readjusting the voltage without interrupting the load current is required. One of the most economical ways of doing this is to provide taps on the low side of a transformer and an automatically controlled switch that will change the tap connections as required to maintain a constant output voltage. A typical schematic is shown in Figure 3-30. For simplicity only three taps (labeled *a*, *b*, and *c*) have been shown. The arrows labeled (*f*) and (*g*) are switch contacts or fingers that can move to any of points *a*, *b*, and *c*. The reactor coil (*d*) is center tapped, with sufficient turns of wire to withstand the voltage between adjacent taps and large enough wire to carry half the load current on each side of the center

tap. The switching operations are as follows. With (*f*) and (*g*) both on point (*a*), the output voltage is at a minimum and half of the load current flows in each side of the reactor coil as shown. But if (*f*) remains on (*a*) and (*g*) is moved to (*b*), the reactor (*d*) will serve as a voltage divider, and the output voltage will be increased. A magnetizing current will now flow in the reactor, and half the load current also flows in each side of the reactor coil. If (*f*) is now moved to (*b*) and (*g*) remains on (*b*), a further increase of output voltage will occur. Moving (*g*) to (*c*) will again raise the voltage, and moving (*f*) to (*c*) will give the maximum output voltage. Additional taps on the transformer winding would provide more voltage steps but the switching is always such that (*f*) and (*g*) are never separated by more than one tap. The movement of the contact fingers (*f*) and (*g*) is as rapid as possible, but during the switching operation, all the load current is carried by one half of the reactor coil. The voltage drop across the reactor coil in this condition is considerable, but it can be minimized by designing the reactor to have a rather high magnetizing current (about 60% of the full-load current).

For very large regulating transformers, it is preferable to use a tap-changing arrangement on the main transformer and a series transformer through which the tapped voltage can be added to (or subtracted from) the output voltage. These are known as two-core regulating transformers. Figure 3-31 shows one such possible arrangement. Use of a series transformer reduces the switching duty imposed on the tap-changing switches.

If a voltage regulator is required but voltage transformation is not, the regulating transformers shown in Figures 3-30 and 3-31 can be reduced to autotransformer designs, such as those shown in Figure 3-32, with a substantial reduction of cost.

FIGURE 3-31   Elementary form of two-core
regulating transformer

SERIES TRANSFORMER

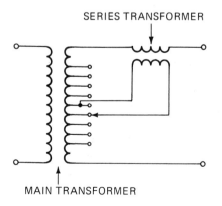

MAIN TRANSFORMER

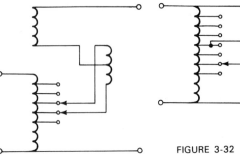

FIGURE 3-32   Two tap-changing voltage regulators

MATHEMATICAL INTRODUCTION TO TRANSFORMERS

Fundamental Transformer Equation

The most important single equation related to transformer design is probably that relating the number of turns of wire required to the voltage, the frequency, the cross section of the magnetic circuit, and the maximum permissible flux density. The relationship is derived as follows. For simplicity we will consider a core-type transformer, but the equations are equally valid for other designs.

The first elementary relationship is given by Equation 3–24.

$$\phi = BA \qquad (3-24)$$

where ϕ = flux, in webers
B = flux density, in teslas
A = cross-sectional area of the flux path, in square meters

The magnetic characteristics of iron are such that the absolute flux density (B) is more or less limited by saturation, so ϕ for any given core also has a more or less definite ceiling. We shall assume that in its cyclic variations ϕ goes to equal positive and negative values. For example,

$$|\phi_{max}| > \phi > -|\phi_{max}| \qquad (3-25)$$

Equation 3–25 is reasonably accurate for a transformer under steady-state conditions. Consider a coil of wire on a core where the flux goes through one cycle of variation from zero to $+\phi_{max}$, to zero, to $-\phi_{max}$, and back to zero. Each turn of the coil will be cut by $4\phi_{max}$ webers (Wb) per cycle, and so the average voltage induced in the coil will be

$$E_{avg} = 4Nf\phi_{max} \qquad (3-26)$$

where N = number of turns in the coil
f = frequency, in hertz

If the coil resistance is small, the induced voltage and applied voltage become practically equal; and if the applied voltage is sinusoidal, the effective voltage is about 1.11 times the average.[3] We can therefore write Equation 3–27, which is the fundamental transformer equation.

$$V = 4.44Nf\phi_{max} \qquad (3-27)$$

It should be emphasized that Equation 3–26 is valid for any waveshape of applied voltage as long as there are only two zero crossings per cycle, and Equation 3–27 only applies if the voltage has a form factor of 1.11.

[3]1.1107207 is a better approximation.

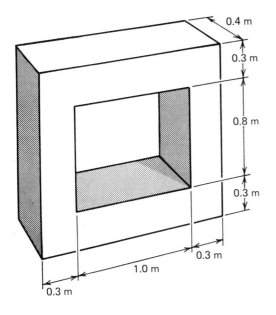

turns required by the transformer core can be found. One method is the same as that used for cores that are magnetized with dc current, with an additional step to convert the calculated peak current to an rms value.

EXAMPLE 3-2

If the transformer core in Example 3-1 has the magnetic characteristics of sheet steel shown in Figure 3-33, find the no-load current on the 14,400-V winding.

Solution

From the dc magnetization curve in Figure 3-33, we require about 1000 ampere turns of magnetizing force per meter of magnetic path. From the sketch of the core, the mean length of the circuit is

$$0.8 + 0.3 + 1.0 + 0.3 = 2.4 \text{ m}$$

$$\therefore I_{max} \text{ at no load} = \frac{1000 \times 2.4}{322}$$

$$= 7.453 \text{ A}$$

and

$$I_{rms} \text{ at no load} = 7.453 \times 0.7071$$

$$= 5.27 \text{ A}$$

This method of calculating the no-load current is quite accurate for finding the peak value of the no-load current, but multiplying by 0.7071

EXAMPLE 3-1

How many turns of wire are required for a 14,400-V 60-hertz (Hz) primary winding on the core shown above if the maximum flux density is to be 1.4 teslas (T)?

Solution

The cross-sectional area of the magnetic path is $0.3 \times 0.4 = 0.12 \text{ m}^2$.

$$\phi_{max} = 0.12 \times 1.4 = 0.168 \text{ Wb}$$

$$14,400 = 4.44N \times 60 \times 0.168$$

$$N = \frac{14,400}{4.44 \times 60 \times 0.168}$$

$$= 321.75 \text{ turns} \qquad \text{(Answer)}$$

Since a fraction of a turn is impractical, use 322 turns.

Calculation of Magnetizing Current and Core Losses

When the maximum flux density in the core has been determined, the no-load ampere

FIGURE 3-33 Typical B–H curve for silicon alloy sheet steel

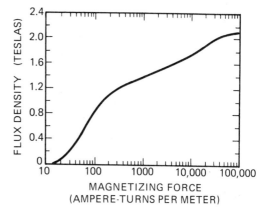

to get the effective value gives pessimistic (high) results (see pp. 87–88).

If the magnetic characteristics of the iron are given in terms of volt-amperes per kilogram (kg), the calculation is equally simple.

Core losses are also calculated from watts loss per kilogram data, but both of these calculations will be subject to some errors created by the specific shape of the core, the existence of bolt holes in the core, burring of the edges of the laminations, and other factors.

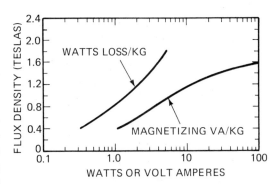

FIGURE 3-34   Typical magnetizing volt-amperes and watts loss per kilogram for silicon alloy sheet steel

EXAMPLE 3-3

If the transformer core in Example 3-2 has magnetic characteristics given by the graph in Figure 3-34, and the core weighs 2016 kg, find the no-load current and the total core losses.

Solution

From Figure 3-34, a flux density of 1.4 T requires about 28 VA/kg.

$$\therefore I_{NL} = \frac{2016 \times 18}{14{,}400} = 3.92 \text{ A} \quad \text{(Answer)}$$

From Figure 3-34, core loss will be about 3.0 W/kg; therefore, total core losses are

$$2016 \times 3.0 = 6048 \text{ W} \quad \text{(Answer)}$$

Effects of Voltage and Frequency Variations

Power transformers are not ordinarily subject to frequency variations and usually are subject to only modest voltage variations, but it is interesting to consider the effects thereof. We shall illustrate the effects by means of examples.

EXAMPLE 3-4

If the transformer from Example 3-1 is operated at 12 kV, 50 Hz, find the maximum flux density in the core.

Solution

We can do this two ways:

a. Working from the fundamental equation,

$$12{,}000 = 4.44 \times 322 \times 50$$
$$\times \phi_{max}$$

$$\phi_{max} = \frac{12{,}000}{4.44 \times 322 \times 50}$$

$$= 0.16787 \text{ Wb}$$

$$\therefore B = \frac{0.16787}{0.12}$$

$$= 1.3989 \text{ T} \quad \text{(Answer)}$$

b. Working on a ratio and proportion idea,

$$\frac{14{,}400}{12{,}000} = \frac{4.44 \times \cancel{N} \times 60 \times \cancel{0.3} \times \cancel{0.4} \times 1.4}{4.44 \times \cancel{N} \times 50 \times \cancel{0.3} \times \cancel{0.4} \times B}$$

$$B = \frac{12{,}000 \times 60}{14{,}400 \times 50} \times 1.4$$

$$= 1.4 \text{ T} \quad \text{(Answer)}$$

The difference between these answers is due to the rounded-off number of turns used in the first method. Since the frequency and voltage changed in the same proportion, the flux density remained the same, and it is immediately apparent that the no-load current will be unchanged.

Changing the voltage and/or frequency also affects the iron losses in a transformer. As long as the flux variations are sinusoidal with respect to time, hysteresis loss (P_h) and eddy current losses (P_e) follow Equations 3-28 and 3-29, respectively.

$$P_h = K_h f(\phi_{max})^x \qquad (3\text{-}28)$$

The exponent x is about 1.4 to 1.8, depending on the grade of iron. We will use 1.8 in our examples.

$$P_e = K_e f^2(\phi_{max})^2 \qquad (3\text{-}29)$$

EXAMPLE 3-5

A transformer rated at 120 V, 60 Hz has 10-W hysteresis loss and 15-W eddy current loss under these conditions. Find the following:

a. The eddy current and hysteresis losses when operated at 120 V, 400 Hz.
b. The voltage required to produce normal flux density in the core at 400 Hz and the value of the individual core losses under those conditions.

Solution

a. With a constant applied voltage, the flux becomes inversely proportional to the frequency; in this case, 60/400 = 0.15 times normal. So by ratio and proportion,

$$\text{Hysteresis loss} = 10 \times \frac{400}{60} \times 0.15^{1.8}$$

$$= 3.873 \text{ W}$$

$$\text{Eddy current loss} = 15 \times \left(\frac{400}{60}\right)^2 \times 0.15^2$$

$$= 15 \text{ W} \qquad \text{(Answer)}$$

b. To get the same flux in the core, the voltage must change in direct proportion to the frequency.

$$\therefore \frac{120}{60} = \frac{V}{400}$$

$$V = \frac{120 \times 400}{60} = 800 \text{ V}$$

If the flux is unchanged,

$$\text{New } P_h = 10 \times \frac{400}{60}$$

$$= 66.66 \text{ W}$$

$$\text{New } P_e = 15 \times \frac{400^2}{60}$$

$$= 666.67 \text{ W} \quad \text{(Answer)}$$

From Examples 3-4 and 3-5 it is apparent that a transformer can be operated at less than rated frequency if the voltage is correspondingly reduced, but raising the voltage and frequency in the same proportion may raise the core losses to an intolerable level. Raising the frequency and holding the voltage constant will reduce the hysteresis loss and leave the eddy current loss unchanged. Some increase of voltage could therefore be tolerated at higher frequencies, but exactly how much depends on the relative magnitude of the hysteresis and eddy current losses and the magnitude of the exponent of the flux term in the hysteresis equation (Equation 3-28).

No-Load Current Wave Shape

With a sinusoidal applied voltage (V_1), the counter emf (E_1) in a transformer has to be a sinusoid, and since E_1 is proportional to the rate of change of ϕ, the latter must vary cosinusoidally with respect to time (e.g., a 90° phase difference must exist between ϕ and E_1). Because of hysteresis the flux will slightly lag the current, and so the current lags the voltage by an angle less than 90°. Eddy current losses also have the effect of decreasing the angle between the voltage and the no-load current. The no-load current can therefore be regarded as having a magnetizing component that is 90° behind the applied voltage

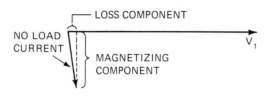

FIGURE 3-35   Phasor representation of
transformer no-load current

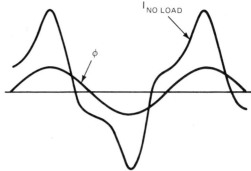

FIGURE 3-36   No-load current of a transformer

and a loss component that is in phase with the voltage, as shown in Figure 3-35.

Since the flux must vary cosinusoidally and the iron has nonlinear *B–H* characteristics (requires very high *H* to get moderate increases of *B* near saturation), the magnetizing current is not altogether sinusoidal but instead has a pointed wave form, as shown in Figure 3-36. The result is that the rms value of the no-load current will be about 0.5 to 0.6 times the peak current.

As shown in Figure 3-37, the no-load current to a transformer can be approximated by the sum of a fundamental frequency current plus a triple frequency current added together. Other frequencies exist, but the fundamental and the third harmonic are the predominant components.

Equivalent Circuit for Power Frequency Transformers

An equivalent circuit for an electrical machine is a circuit composed of linear resistors, inductors, and capacitors (and possibly some idealized voltage sources) from which we can calculate the performance of the machine; it usually bears little physical resemblance to the machine itself. A theoretically exact equivalent circuit permits calculation of the actual machine performance under any and all conditions, but in many cases it is neither necessary nor desirable to use an exact equivalent circuit for each and every calculation.

For example, a practical transformer has small amounts of capacitance between consecutive turns of wire in each coil, some capacitance between the primary and secondary windings, and capacitance between each winding and ground. These capacitances are significant at high frequencies and have considerable effect on the way the transformer responds to an impulse (such as a lightning stroke), but they are insignificant at power frequencies and can be neglected for steady-state calculations.

With regard to a transformer, it is not possible to represent it exactly by means of fixed linear components because of the non-

FIGURE 3-37   Fundamental and third-harmonic
components of no-load current

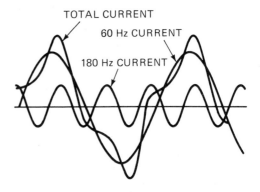

linear magnetic characteristics of the core. There is no way that linear impedances can account for the harmonic components of the no-load current, and even the fundamental component of the no-load current and the iron losses can only be accounted for by impedances that change somewhat with respect to voltage and frequency; such circuits are difficult to solve. So we generally use a circuit with fixed impedances that can be solved using conventional techniques and will give results that are accurate enough for the problem at hand.

Along with the question of accuracy there is also the matter of circuit complexity. The time required to solve an exact equivalent circuit may often be prohibitive, and so various approximate equivalent circuits are also commonly used. For design purposes a transformer is best represented by an equivalent π circuit, but an equivalent T circuit can also be used, and the latter is much more widely known. However, for many purposes (particularly from the user's standpoint) these circuits are unnecessarily complex and will not be treated here. It is easier to view a practical transformer as an ideal transformer to which has been added a parallel impedance on the primary side (to account for iron losses and no-load current) and a series impedance on the secondary side (to account for copper losses and voltage regulation), as shown in Figure 3-38. It should be emphasized that X_e and R_e in Figure 3-38 account for the resistance of both the primary and secondary windings.

Reflected Impedance

Before we can derive values for the equivalent circuit in Figure 3-39, we must understand how an impedance on one side of a

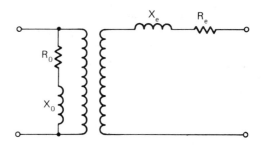

FIGURE 3-38 Equivalent circuit for a transformer

transformer can be replaced by an equivalent impedance on the other side of the transformer. The meaning of the word "equivalent" and the steps required to find its value are illustrated in Figure 3-39 and the subsequent explanation.

If an ideal transformer having a turns ratio of N_1/N_2 has a connected load Z_L, the current in the secondary is simply V_2/Z_L. The primary voltage is just $(a \cdot V_2)$, and the primary current is $[(V_2/Z_L) \div a]$.

FIGURE 3-39 Equivalent or reflected impedance

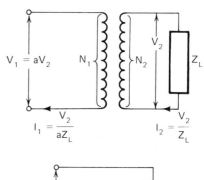

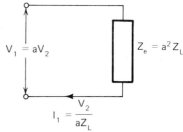

The reflected or equivalent impedance is that which if directly connected to the primary circuit would permit the same primary current flow, and is therefore $aV_2/[(V_2/Z_1) \div a] = a^2 Z_L$. In other words, to find an equivalent impedance on the other side of a transformer, multiply the known impedance by the square of the turns ratio (or its inverse, depending on where the original impedance is located).

Measurement of Losses and Calculation of Efficiency

Because of its very high efficiency, it is not practical to determine the efficiency of a transformer by direct measurement of the input and output power under load. Even if the wattmeters were accurate to within $\pm\frac{1}{2}$ of 1%, direct measurement could indicate an efficiency of over 100% for a transformer that is actually over 99% efficient. So the usual procedure is to measure the losses and calculate the efficiency for various load conditions.

The iron losses of a transformer can be measured quite accurately by simply measuring the power input at rated voltage and frequency and no load. Theoretically, it makes no difference which winding is energized, but it is usually more convenient to energize the low side. The current during this open-circuit test will be quite small, so the primary copper losses will be small[4] (we will neglect them), and the secondary copper losses will be zero. It is important to use rated voltage for this test, and although not required for efficiency calculations, we generally record the current observed during the test.

[4] The primary copper loss can be calculated from current and resistance measurements and subtracted from the no-load power input to get a more accurate value for the iron losses.

The copper loss can be measured by short circuiting the transformer and measuring the power input at rated frequency and full load current.[5] It is usually convenient to perform this short-circuit test by shorting out the low-voltage winding and energizing the high side, but theoretically it does not matter which way the test is done. The voltage during this test is normally quite small (less than 10% of the rated value), so the iron losses are small and we will neglect them. It is very important that the current used during the short-circuit test be recorded, and although it is not required for efficiency calculations, we generally record the voltage as well.

We generally use the circuit in Figure 3-38 to calculate the efficiency (e.g., the iron losses are considered to be constant). From the short-circuit test we can determine R_e and then use that value to find the copper loss at any value of load current. However, we usually neglect the effect of voltage regulation so that kVA output and current are directly proportional to each other. Using that approximation, the efficiency can be calculated, using Equation 3–30 (see top of facing page), without finding R_e or R_o.

EXAMPLE 3-6

A certain 3-kVA transformer has 20-W iron loss at rated voltage and 50-W copper loss at full load. Find the efficiency with a 2-kVA, 65% pf load.

Solution

$$\eta = \frac{2000 \times 0.65}{2000 \times 0.65 + 20 + 50\left(\frac{2}{3}\right)^2}$$

[5] Usually this test is done with the transformer at ambient temperature (T_1), and the measured losses should be corrected to the temperature at which the unit will actually operate (T_2).

$$P_{T_2} = P_{T_1}\left(\frac{T_2 + 234.5}{T_1 + 234.5}\right)$$

$$\eta = \frac{\text{VA load} \times \text{pf}}{\text{VA load} \times \text{pf} + P_i + P_{cu}\left(\dfrac{\text{kVA load}}{\text{kVA rating}}\right)^2} \qquad (3\text{-}30)$$

where P_i = iron loss at rated voltage, in watts

P_{cu} = copper loss at full load, in watts

pf = load power factor

$$= \frac{1300}{1300 + 20 + 22.22}$$

$$= 0.9685 \qquad \text{(Answer)}$$

We can learn several things by examining Equation 3-30. First, changing the load power factor does not change the losses, so raising the load power factor will improve the efficiency of a transformer because the losses then become a smaller proportion of the total power input. Second, the no-load efficiency of a transformer (or any other machine) must be zero, because the numerator of Equation 3-30 goes to zero while the denominator never will. Third, the rapid increase of copper losses at high loads will decrease the efficiency under that condition. It therefore also follows that maximum efficiency will occur at some intermediate value of load.

Students with a calculus background can differentiate the right-hand side of Equation 3-30, set it equal to zero, and thereby derive Equation 3-31, which states that maximum efficiency occurs when the copper losses become equal to the iron losses.

For η (max):

$$P_i = P_{cu}\left(\frac{\text{kVA load}}{\text{kVA rating}}\right)^2 \qquad (3\text{-}31)$$

EXAMPLE 3-7

Referring to the transformer in Example 3-6, find the following:

a. The load level at which maximum efficiency occurs.
b. The maximum efficiency with (1) a 100% pf load; (2) a 40% pf load.

Solution

a. From Equation 3-31,

$$20 = 50\left(\frac{\text{kVA load}}{3}\right)^2$$

$$\text{kVA load} = \sqrt{\frac{20 \times 3^2}{50}} = 1.897 \quad \text{(Answer)}$$

b. With a 1.897-kVA load, the copper loss equals the iron loss for a total of 40 W. Since efficiency = output/(output + losses),

$$(1) \ \eta = \frac{1897}{1897 + 40}$$

$$= 0.97935 \qquad \text{(Answer)}$$

$$(2) \ \eta = \frac{1897 \times 0.4}{(1897 \times 0.4) + 40}$$

$$= 0.94992 \qquad \text{(Answer)}$$

Another problem that crops up is the calculation of all-day efficiency if the load cycle and transformer losses are known. The easiest approach is to sum up the energy used by the load, and also find the energy used by the losses. The all-day efficiency is just the ratio (energy out)/(energy in).

EXAMPLE 3-8

The transformer from Example 3-7 operates for 6 hours at full load, 100% pf; 4 hours at half load, 80% pf; and no load the balance of a 24-hour day. Find its all-day efficiency.

Solution

Energy out = (6 × 3000 × 1.0)
              + (4 × 1500 × 0.8)

           = 22,800 watt-hours (Wh)

Energy lost:

Through iron losses
           = 24 × 20 = 480 Wh

Through copper losses
           = (6 × 50) + [4 × 50 × ($\frac{1}{2}$)^2]

           = 350 Wh

All-day efficiency

   = $\dfrac{22,800}{22,800 + 480 + 350}$

   = 0.96488   (Answer)

A fourth kind of problem which comes up is that of determining the minimum size of transformer required to feed a load that goes through some definite cycle. If the load has a rather long cycle (e.g., several hours), the thermal time constant of the transformer must be considered, and the solution of such a question is beyond the scope of this book. But if the load cycle is short enough that its temperature will not change appreciably during the one cycle, the minimum transformer size is the rms value of the load, which can be found by drawing a graph of the load cycle like that in Figure 3-40 and applying Equation 3–32.

$$S_{rms} = \sqrt{\frac{S_1^2 t_1 + S_2^2 t_2 + \cdots + S_n^2 t_n}{t_1 + t_2 + \cdots + t_n}} \quad (3\text{–}32)$$

EXAMPLE 3-9

A cyclic load draws 100 kVA for 1 min, 60 kVA for 3 min, and nothing for the balance of its 10-min cycle. Find the minimum size of transformer required (see figure at right).

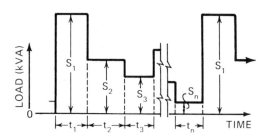

FIGURE 3-40   Graph of a load cycle

Solution

$$S_{rms} = \sqrt{\frac{(100^2 \times 1) + (60^2 \times 3) + (0^2 \times 6)}{10}}$$

   = 45.6 kVA   (Answer)

When applying a transformer on the basis of rms load, it is important to see that the voltage regulation at peak load will not be excessive.

There is also the question of how well a transformer will withstand short-circuit currents of extremely short duration. Short-circuit currents are destructive in two ways, through mechanical forces and the heating effects. The manufacturer usually braces the windings for the maximum short-circuit current that can flow through the transformer with rated voltage maintained on the primary and a short circuit applied right at the secondary terminals, so the mechanical forces are usually not a point of concern. How-

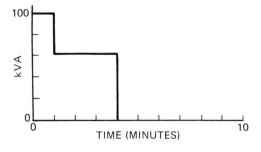

ever, an exception to this is a mutual-induction transformer used as an autotransformer, and this problem is treated on pp. 95–96. The thermal problem is basically a matter of how much heat can be stored in the windings before an objectionable temperature is reached. Since heat (in joules) is just $|I|^2 Rt$, and R is essentially constant for a given unit, the transformer should be regarded as having a definite $|I|^2 t$ limitation. Example 3-10 shows an easy way to deal with this problem, if the maximum $|I|^2 t$ value is known, but this method is only valid for rather small values of t, perhaps 10 s or so.

EXAMPLE 3-10

The manufacturer of a certain transformer states that it can carry 25 times its rated current for 1 s.

a. For how long can a current of 15 times the rating be allowed to flow?
b. How much current could the transformer carry for 2 s?

Solution

It is convenient and perfectly valid to work in terms of the transformer rating. Based upon equal $|I|^2 t$,

a. $25^2 \times 1 = 15^2 \times t$

$t = 2.778\ s$ (Answer)

b. $25^2 \times 1 = I^2 \times 2$

$I = 17.68$ times full-load current   (Answer)

Voltage Regulation and Impedance Calculations

Numerical values for R_o and X_o in the equivalent circuit of Figure 3-41 can be found from the voltage, current, and power measurements obtained during the open-circuit test. The procedure is outlined by Equations 3–33 through 3–35.

$$|Z| = \frac{|V|}{|I|} \tag{3-33}$$

$$|R| = \frac{P}{|I|^2} \tag{3-34}$$

$$|X| = \sqrt{|Z|^2 - |R|^2} \tag{3-35}$$

If the measurements were obtained on the low side of the transformer, the ohmic values obtained will be equivalent low-side values, but if high-side values are required, multiply by the square of the turns ratio. However, R_o and X_o do not enter into voltage regulation or impedance calculations, so their values (and the test required to obtain them) are not required.

In similar fashion, numerical values for R_e and X_e in the equivalent circuit of Figure 3-38 can be found from the voltage, current, and power measurements obtained during the short-circuit test. Work from the actual measurements, apply temperature corrections to the resistance values as necessary, and convert the ohmic values to equivalent "other side" values if required. (See Example 3-11.)

FIGURE 3-41   Schematic and phasor diagram for voltage-regulation calculations

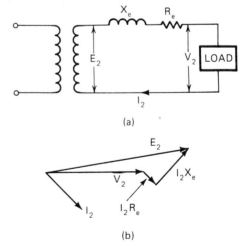

(a)

(b)

EXAMPLE 3-11

Transformer rating:
   25 kVA, 2400/120 V; temperature rise at
   full load = 75°

Short-circuit test
   $V = 65$ V $\}$
   $I = 10.4$ A $\}$ high-side measurement
   $P_{cu} = 280$ W $\}$

Open circuit test:
   $V = 120$ V $\}$
   $I = 0.7$ A $\}$ low-side measurement
   $P_i = 45$ W $\}$

Temperature of the transformer during the
   test = 20°C

Find the values of R_o and X_o on the high side
and R_e and X_e on the low side assuming that
the transformer will operate in a 40°C ambient.

Solution

$$|Z_o| = \frac{120}{0.7} \times 20^2$$

$$= 68,572 \ \Omega$$

$$R_o = \frac{45}{0.7^2} \times 20^2$$

$$= 36,736 \ \Omega$$

$$|X_o| = \sqrt{68,572^2 - 36,736^2}$$

$$= 57,900 \ \Omega$$

$$|Z_e| = \frac{65}{10.4} \div 20^2$$

$$= 0.015625 \ \Omega$$

(Cold) $R_e = \dfrac{280}{10.4^2} \div 20^2$

$$= 0.006473 \ \Omega$$

$$|X_e| = 0.015625^2 - 0.006473^2$$

$$= 0.01422 \ \Omega$$

(Hot) $R_e = 0.006473 \times \left(\dfrac{115 + 234.5}{20 + 234.5}\right)$

$$= 0.008889 \ \Omega$$

If full-load current is used during the short-circuit test, the voltage observed will be the impedance voltage of the tranformer, but if temperature corrections are to be applied, use X_e and R_e (hot) to obtain $|Z_e|$, and find $|IZ|\%$ using Equation 3–36.

$$|IZ|\% = \frac{|I_2| \times |Z_e|}{|V_2|(\text{rated})} \qquad (3\text{–}36)$$

where $|I_2|$ is rated secondary current.

To deal with voltage-regulation questions, the reader is referred to the schematic and phasor diagram in Figure 3-41. Under no-load condition, $|V_2|$ becomes equal to $|E_2|$, and the voltage regulation is just the difference between these absolute quantities. The phasor diagram in Figure 3-41 is constructed as follows. Draw $|V_2|$ at the 0° position and draw $|I_2|$ at the appropriate[6] position therefrom. If $|I_2|$ and $|R_e|$ are known, $|I_2 R_e|$ can be determined and drawn parallel to I_2. If $|X_e|$ is known, $|I_2 X_e|$ can be similarly found and drawn 90° ahead of $|I_2|$. The no-load voltage $|E_2|$ is just the phasor sum of $|V_2| + |I_2 R_e| + |I_2 X_e|$.

If all other factors are known, $|E_2|$ in the phasor diagram can be readily found. But if the problem is to find $|V_2|$ when all other factors are known, two algebraic solutions are available.

The first solution is to split each phasor into its x and y components and use the Pythagorean theorem. If we do so, we get Equation 3–37.

___

[6]Cos $\angle{}_{I_2}^{V_2}$ is the load power factor; so if the latter is specified, the position of I_2 can be readily determined. If the load power factor is not known for some particular problem, draw I_2 in an approximate position, complete the phasor diagram in rough form, and determine the exact position of I_2 later.

$$E_2 = \sqrt{(|V_2| + |I_2R_e|\cos\phi + |I_2X_e|\sin\phi)^2 + (|I_2X_e|\cos\phi - |I_2R_e|\sin\phi)^2} \qquad (3\text{-}37)$$

where ϕ is considered to be positive for lagging power factor loads.

The second solution is to split $|V_2|$ into two components, one in phase with I_2 and the other 90° from it. We can then write Equation 3-38.

$$E_2 = \sqrt{(|V_2|\cos\phi + |I_2R_e|)^2 + (|V_2|\sin\phi + |I_2X_2|)^2} \qquad (3\text{-}38)$$

where ϕ is considered to be positive for lagging power factor loads.

EXAMPLE 3-12

If the transformer in Example 3-11 has a no-load voltage of 122 V and is supplying a 208-A, 0.65-pf lagging load, find its voltage regulation in percent.

Solution

Since pf = cos ϕ = 0.65,

$$\phi = 49.46°$$

$$\sin\phi = 0.7599$$

Using Equation 3-37

$$122 = \{|V_2| + (208 \times 0.008889 \times 0.65)$$
$$+ (208 \times 0.01422 \times 0.7599)^2$$
$$+ (208 \times 0.01422 \times 0.65)$$
$$- (2.08 \times 0.008889 \times 0.7599)^2\}^{1/2}$$

$$122 = \{(|V_2| + 1.202 + 2.248)^2$$
$$+ (1.923 - 1.405)^2\}^{1/2}$$

$$122^2 = (|V_2| + 3.45)^2 + 0.518^2$$

$$(|V_2| + 3.45)^2 = 14{,}884$$

$$|V_2| + 3.45 = 121.999$$

$$|V_2| = 118.549 \text{ V}$$

$$\text{Voltage regulation} = \frac{122 - 118.549}{118.549} \times 100$$

$$= 2.911\% \qquad \text{(Answer)}$$

Equations 3-37 and 3-38 are exact, but Equations 3-39 and 3-40 are useful approximations.

$$|E_2| = |V_2| + |I_2R_e|\cos\phi + |I_2X_e|\sin\phi$$
$$+ \frac{1}{|V_2|}(|I_2X_e|\cos\phi - |I_2R_e|\sin\phi)^2 \qquad (3\text{-}39)$$

$$|E_2| = |V_2| + |I_2R_e|\cos\phi$$
$$+ |I_2X_e|\sin\phi \qquad (3\text{-}40)$$

Performance of Mutual-Induction Transformers Used As Autotransformers

When a mutual-induction transformer is connected as an autotransformer, it is convenient to view its impedance as being entirely in the coil that is used as the extension winding, as shown in Figure 3-42. If the voltage per turn is the same, the copper losses

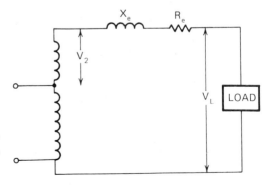

FIGURE 3-42 Mutual-induction transformer used as an autotransformer

remain constant no matter how the transformer is being used.

If the kVA rating as a mutual-induction transformer is known, the capacity as an autotransformer is given by Equation 3–41.

kVA capacity as auto

$$= \text{kVA rating} \times \frac{|V_L|}{|V_2|} \quad \text{(3–41)}$$

If the copper and core losses as a mutual induction transformer are known and they remain unchanged, the efficiency can be calculated using Equation 3–42.

$$\eta = \frac{P(\text{load})}{P(\text{load}) + \text{total losses}} \quad \text{(3–42)}$$

If the impedance (in ohms) does not change, the new impedance (in percent terms) is given by Equation 3–43.

(New) $|Z\%|$

$$= (\text{original}) |Z\%| \times \frac{|V_2|}{|V_L|} \quad \text{(3–43)}$$

Equations 3–41 through 3–43 are valid no matter whether a buck or a boost connection is used, and they show why the following are true:

■ The efficiency of an autotransformer is extremely high.
■ The voltage regulation is very low.
■ The short-circuit current is very high.
■ The physical size of an autotransformer is very small considering the load it can supply.

Phasor Diagrams for Transformers and Three-Phase Banks

There is a wide range of problems related to transformers that requires the use of phasor diagrams for their solution. We shall therefore present the rules to be followed

when constructing such phasor diagrams and do a few examples.

Unless otherwise specified, it is convenient to neglect the load currents of the transformers and also to neglect the voltage regulation. We will use the following conventions and double subscript notation.

■ *Rule 1:* $V_{AB} = E_{BA} = -E_{AB} = -V_{BA}$.
■ *Rule 2:* $V_{AN} = V_{AB} + V_{BC} + V_{CD} + \cdots + V_{MN}$, where A, B, C, and so on, are points along any continuous path in the circuit.

In conjunction with single-phase transformers, we will designate their terminals in the usual fashion (H_1, H_2, X_1, X_2, and so on), and we have the following:

■ *Rule 3:* $V_{H1\text{-}H2}$ is in phase with $V_{X1\text{-}X2}$.
■ *Rule 4:* $I_{H1\text{-}H2}$ is in phase with $I_{X2\text{-}X1}$.
■ *Rule 5:* For a center-tapped coil (X_2 being the center tap), $V_{X1\text{-}X2} = V_{X2\text{-}X3} = \frac{1}{2}V_{X1\text{-}X3}$.

When dealing with transformer banks, we will identify the individual transformers with letters (like P, Q, and R) and write expressions like $V_{QX1\text{-}X2}$, which means "external voltage on transformer Q in the direction X_1 to X_2," or $E_{PH2\text{-}H1}$, which is "the internal voltage of transformer (P) in the direction H_2 to H_1." Let us do two examples.

EXAMPLE 3-13

Referring to the following diagram and assuming the primary phase sequence is 1, 2, 3, find the phase sequence of the secondary voltage and determine whether the secondary line voltages lead or lag the primary line voltages.

Solution

Start by drawing $V_{1\text{-}2}$ at the 0° position. From the phase sequence quoted, $V_{2\text{-}3}$ must fall at −120° and $V_{3\text{-}1}$ must be at +120°, so we can draw them also. Since $V_{1\text{-}2} = V_{PH1\text{-}H2}$, and $V_{PX1\text{-}X2}$ is synonymous with V_{AN}, the latter is shown coincident with (but smaller than) $V_{1\text{-}2}$.

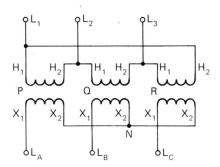

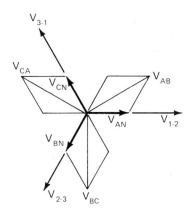

V_{BN} and V_{CN} are located using similar arguments. Since $V_{AB} = V_{AN} + V_{NB}$, and so on, we can now draw V_{AB} and the other two line voltages. If we now stop and look at the phasor diagram, it is apparent that the secondary phase sequence is *ABC*, and the secondary line voltages lead the primary voltages by 30°.

EXAMPLE 3-14

In the following diagram, transformer *P* is center tapped on both the high and low sides, and the primary voltage is 2400 V, balanced three phase. Find the voltage across the high side of transformer *Q*.

SUMMARY

The transformer is a very important electrical machine; without it, power systems as we know them today could not exist.

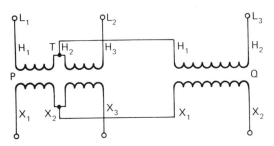

Solution

Since the phase sequence is not specified, we will draw V_{1-2} at 0°, V_{2-3} at –120°, and V_{3-1} at +120°. We can see from the schematic that $V_{1-3} = V_{1-T} + V_{T-3}$, and so $V_{T-3} = V_{QH1-H2} = V_{1-3} - V_{1-T}$. We can also see that $V_{1-T} = \frac{1}{2}V_{1-2} = 1200 \angle 0°$. So $V_{T-2} = 2400 \angle -60° - 1200 \angle 0° = 2078 \angle -90°$ V. The complete phasor diagram is shown below.

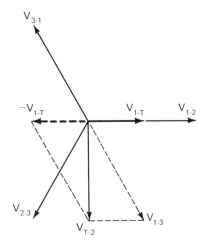

The transformer connection shown in Example 3-14 is the well-known T connection. It is quite commonly used for three-phase transformer assemblies and is a good example of how phasors can be applied.

Because thousands of transformers are manufactured every year, their design and construction have been refined to a very high degree; but their theory of operation is quite simple and can be readily visualized in terms of physical concepts or expressed mathematically with a high degree of accuracy. The most difficult thing about the mathematical treatment is probably the large number of alternative ways in which most problems can be attacked. But their phasor diagrams are perfectly straightforward, and the mathematical operations are not particularly difficult.

Not all possible questions about transformers have been answered in this book. We have provided an introduction to transformer design, but a designer needs considerably more information than has been included here.

However, people who install, repair, or select and apply transformers will find that most of the information they require has been presented.

REVIEW QUESTIONS

3-1. Explain why the voltage ratio of a transformer is always approximately equal to the turns ratio.

3-2. Explain why the current ratio in a two-winding transformer is approximately the inverse of the turns ratio.

3-3. A transformer does not necessarily constitute a lagging power factor load on the supply system. Explain why.

3-4. With reference to transformers, explain what is meant by the following terms:
   (a) Primary
   (b) Secondary
   (c) Copper losses
   (d) Hysteresis loss
   (e) Eddy current loss
   (f) Efficiency
   (g) Voltage regulation
   (h) Leakage flux
   (i) Impedance voltage
   (j) Polarity

3-5. What is the usual range of transformer full-load efficiency and how is it affected by load power factor?

3-6. Why does the cooling problem become progressively more difficult as transformer size increases?

3-7. By means of a sketch and a written explanation, show how to determine the identity of the terminals of a distribution transformer from which the markings have been obliterated.

3-8. What are the four requirements for parallel operation of transformers?

3-9. Explain how to phase out transformers for parallel operation.

3-10. What distinguishes an autotransformer from a mutual-induction type?

3-11. What is the difference between a buck and a boost autotransformer?

3-12. Can a bucking autotransformer be used to step up a voltage? Explain.

3-13. What is meant by neutral instability of a wye–wye transformer bank?

3-14. Why are high ratio step-down autotransformers rarely if ever used?

3-15. If one transformer in a delta–delta bank becomes open circuited, what effect will this have on the output voltages?

3-16. If a switchboard voltmeter is calibrated 0 to 15 kV, what voltage should be expected at its terminals?

3-17. What is the difference between the way the potential and current transformers are connected into a power circuit?

3-18. Why are current transformers always

equipped with a convenient way to make it possible to short circuit the secondary terminals?

3-19. What is meant by ratio and phase-angle errors of instrument transformers and what factors affect them?

3-20. A current transformer is marked "accuracy class 0.3 B–0.5." What is meant by this marking?

3-21. What is a doughnut type of current transformer?

3-22. Why must the primary and secondary windings be kept as close as possible to each other?

3-23. Why are cylindrical coils preferable to rectangular coils for large transformers?

3-24. Why is wire of square cross section preferable to round wire?

3-25. What is the difference between a shell- and a core-type transformer?

3-26. What is meant by a wound core?

3-27. What is distinctive about the magnetic characteristics of grain-oriented silicon steel?

3-28. Draw schematic diagrams of transformers that have the following voltage ratings and show the standard terminal markings;
(a) 2400/120–240 V.
(b) 2400/240–120 V.
(c) 2400/120 × 240 V.

3-29. Referring to question 3-28, explain the difference between transformers (a) and (c).

3-30. What is meant by a voltage regulator?

3-31. What is meant by "on load tap changing," and why is it required?

MATHEMATICS PROBLEMS

3-1. A transformer rated at 600/208 V has 300 turns on its high-voltage winding. How many turns are there in the low-voltage coil?

3-2. A transformer with a turns ratio of 30 to 1 is supplying 25 A at 240 V. Find the following:
(a) The primary voltage and current.
(b) The volt-amperes input.

3-3. With rated voltage applied to the high winding, a transformer rated at 600/120 V delivered 122 V at no load. If the low winding were energized at 120 V, what no-load voltage would appear on the high winding?

3-4. A transformer with a 4-to-1 turns ratio has 480 V on the high side and delivers 32 A at 0.9 pf on the low side. Find the following:

(a) The high-side current.
(b) The low-side voltage.
(c) The input power (watts).

3-5. If a certain transformer has 97.4% efficiency and the power input is 1500 W, find the power output.

3-6. If the power input to a transformer is 1200 W and the output is 120 V, 19.8 A, at 49% pf, find the efficiency.

3-7. Measurements obtained on the high side of a transformer were 600 V, 8 A at 0.8 pf, and on the low side the measurements were 480 V, 10.1 A and 0.795 pf. If the measurements are exact, which side of the transformer is the primary and what is its efficiency?

3-8. If a 3-kVA transformer is 98% efficient at full load and unity power factor, find the following:

(a) Its losses at full load.

(b) Its approximate efficiency at full load and 65% power factor.

3-9. The physical size of a certain transformer is such that it can dissipate heat at the rate of 15 W. How much load can this unit supply with full-load efficiencies as follows?

(a) 96%

(b) 99%

3-10. A certain transformer has a full-load voltage of 277 V. Find its voltage regulation with no-load voltages as follows:

(a) 281 V

(b) 285 V

(c) 267 V

(d) 277 V

3-11. Find the no-load voltage of the transformer in problem 3-10 if voltage regulations are as follows:

(a) 3%

(b) 5%

(c) −4%

3-12. If the no-load voltage is 245 V and the voltage regulation is 2.6%, find the full-load voltage.

3-13. Find the full-load line current on both sides of each of the following single-phase transformers?

(a) 3 kVA, 2400/240 V (3% *IZ*)

(b) 1 kVA, 240/12 V (2.8% *IZ*)

(c) 25 kVA, 14,400/240 V (3.2% impedance)

(d) 50 kVA, 7200/240–120 V

(e) 40 MVA, 132 kV/25 kV (9.6% impedance)

(f) 150 kVA, 2400 × 4800/240–480 V (5% *IZ*)

3-14. Find the short-circuit current available from each of the transformers in problem 3-13, except (d), if rated voltage is maintained on the high side.

3-15. If a transformer rated at 10-kVA, 240-V output can deliver 1100-A short-circuit current, what voltage regulation will occur if the load power factor is at the most unfavorable value?

3-16. Find the full-load currents on both sides of the following three-phase transformers:

(a) 225 kVA, 4160/480 V

(b) 333 kVA, 2400/208 V

(c) 160 MVA, 240 kV/138 kV

3-17. A set of current, voltage, and power measurements is to be taken in a single-phase circuit using a 14.4-kV, 120-V potential transformer, a 60/5-A current transformer, and meters that are calibrated to directly indicate the voltage (or current or power) at their own terminals. The measurements obtained are 118 V, 4.9 A, and 70 W. Find the voltage, current, and true power input to the load.

3-18. If a transformer rated at 2400/120 V has two overvoltage taps and three undervoltage taps, all in $2\frac{1}{2}$% steps, with what primary voltages can rated secondary voltage be obtained?

3-19. If the transformer in problem 3-18 delivers 125 V when set on its lowest tap, which tap will provide an output voltage of (or near as possible to) 118 V?

3-20. If a maximum flux density of 1.3 T is desired, design a 600-V primary winding for the core shown. (See top of facing page.)

3-21. Referring to problem 3-20, assume the oxide coating on and air spaces between the laminations reduces the effective area of the magnetic circuit to 94% of its apparent area (e.g., a stacking factor of 0.94).

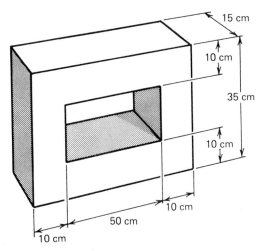

(a) What flux density will be obtained if the number of turns of wire is not changed?

(b) How many turns of wire will be required to maintain a maximum flux density of 1.3 T?

3-22. Referring to problem 3-21(a) and (b), find the no load current for each case based on the following:

(a) DC magnetization curves from Figure 3-33.

(b) AC magnetization curves in Figure 3-34 (the core's total weight is 3.1 kg).

3-23. Referring to problem 3-21(a) and (b), find the total core losses in each case.

3-24. If a certain transformer has 6-W hysteresis loss and 14-W eddy current loss when operated at 120 V, 60 Hz, find the value of these losses when operated at the following:

(a) 100 V, 50 Hz

(b) 200 V, 400 Hz

3-25. If a transformer having a turns ratio of 30 to 1 is connected to a 25-Ω load, find the equivalent load impedance on the high side.

3-26. A transformer rated at 1500 kVA has 3-kW iron loss and 9-kW copper loss at full load:

(a) Calculate the efficiency at 0, 5, 10, 15, 25, 50, 75, 100, and 125% of rated load at 1.0 and 0.6 power factor and graph the results (two graphs).

(b) Find the point of maximum efficiency and the efficiency at that load level for both 1.0 and 0.6 pf.

(c) Find the maximum load this transformer can carry on a 3 min on and 2 min off cycle.

(d) Find the all-day efficiency if the load for part (c) has 0.7 pf and runs only 8 hours/day.

3-27. A certain 100/5-A current transformer has a 1-s rating of 80 times normal.

(a) How many amperes can it carry for 2 s?

(b) For how many seconds can it carry 3000 A?

3-28. The copper losses given for the transformer in problem 3-26 are at a temperature of 115°C. Find the losses that will be measured at 20°C.

3-29. A certain transformer has 0.12-Ω equivalent resistance and 0.35-Ω reactance on its secondary side. When supplying a load of 30 A, 1.0 pf, its terminal voltage is 120 V. Assuming a constant primary voltage, find the following:

(a) The no-load voltage.

(b) The terminal voltage with a 150-A, 0.6-pf lagging load. (Do this and also part c four times using Equations 1–37 through 1–40 and compare the results.)

(c) The terminal voltage with a 200-A, 0.5-pf leading load.

(d) The load power factor at which the voltage regulation will be the greatest, and the amount of regulation that will occur with a 100-A load at that power factor.

(e) The power factor at which a 100-A load will cause zero change of voltage.

(f) The power factor at which a 100-A load causes a 2-V decrease of terminal voltage.

(g) The load current that at 91.4% pf leading will cause no change of terminal voltage.

3-30. A transformer rated at 1 kVA, 240/12 V has 5-m Ω resistance and 7-m Ω reactance on its low side. It is to be used to raise an available line voltage of 240 V to 252 V. The core losses at rated voltage are 25 W.

(a) Draw a diagram to show how the transformer should be connected.

(b) Find the maximum kVA load that can be carried on the 252-V side.

(c) Find the efficiency of the autotransformer at full load, unity pf.

(d) Find the voltage regulation with a 1.0-pf load.

3-31. Referring to Example 3-13, draw a phasor diagram and show that if the primary phase sequence is 1, 3, 2, the secondary phase sequence will be ACB, and the secondary voltages will lag the primary voltage.

3-32. If their output voltages are equal, a wye–delta transformer bank can be paralleled with a delta–wye bank, but the exact details of how to do this are not altogether obvious. Draw a schematic diagram to show how this can be done and a phasor diagram to prove that your schematic is correct.

3-33. Show with phasor diagrams that the T-connected transformers in Example 3-14 can be paralleled with an open delta–open delta bank.

3-34. Referring to Example 3-14, put a tap (N) on transformer Q and determine the position of N that will make $|V_{AN}| = |V_{BN}| = |V_{CN}|$ (i.e., the voltages are indistinguishable from those of a four-wire wye).

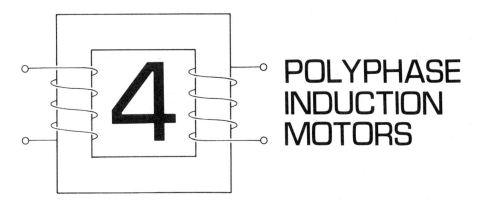

POLYPHASE INDUCTION MOTORS

Like the transformer, the polyphase induction motor helped to make ac power distribution more attractive than dc systems. Direct-current armature windings have to be carefully braced against centrifugal force, commutators are expensive to manufacture, and brush and commutator assemblies require considerable maintenance. In contrast, polyphase induction motors have no commutator, and the most popular (squirrel cage) type has no brushes or slip rings of any kind. The rotor (or rotating part) is practically indestructible, and the insulated windings are on the stationary assembly (called the stator) where centrifugal force does not exist. As a result, these motors are much cheaper and more reliable than dc types.

The term *polyphase* indicates "more than one phase." Many early ac power distribution systems were of two-phase design and so were many of the induction motors used. However, as three-phase power systems gained popularity, so did three-phase mo-

tors, and today there are few two-phase motors in existence. Motors with more than three phases are theoretically possible but do not offer sufficient advantage to make them a commercial success. We will therefore only consider three-phase types.

In the first of this chapter we will develop the theory of induction motor operation, and then we will examine the design of stator windings in more detail and show how dual-voltage motors, part winding and wye–delta motors and two-speed designs are obtained. We will not treat pole amplitude modulation or phase-wound rotors.

The mathematics that follows provides a better grasp of the material in the preceding part of the chapter and can serve as a basis for further study of induction motor design.

Tradesmen and technicians will find that most of the information they need on induction motor theory and windings has been included here. Induction motor characteristics and control are treated in more detail in Chapter 9.

CONSTRUCTION AND THEORY OF OPERATION

Construction of a Squirrel-Cage Motor

A cross-sectional view of the magnetic circuit of a three-phase squirrel-cage motor is shown in Figure 4-1. The stator or stationary core is built up from silicon steel laminations punched and assembled so that it has a number of uniformly spaced identical slots roughly parallel[1] to the armature shaft. In our diagram we have shown 12 slots in the stator core. In practice, there are usually more than 12 slots, integral multiples of 6,

[1]Sometimes the laminations are assembled so that the slots are slightly twisted or skewed. If it exists, the amount of skew will be about equal to the center-to-center distance between adjacent slots.

such as 36, 48, and 72 slots, being quite common.

The slots in Figure 4-1 are semiclosed, and for motors up to 200-kW rating this is the usual arrangement. Larger machines usually have open slots, as shown in Figure 4-2. With semiclosed slots we first line the slots with paper or mylar. Using wire of round cross section, the coils are wound on a form and then inserted into the slots.[2] When the winding is completed, the whole core and winding assembly are dipped in insulating varnish and baked, forming a rigid mass.

With open slots, wire of square or rectangular cross section is used. The coils are wound in precise layers on a form and bent to the exact shape required. These preformed or layer-wound coils are insulated, varnished, and baked, and afterwards are inserted into the slots. In this case the coils must be securely wedged in place to prevent movement, and the coil ends may require a secure mechanical support (such as being tied or lashed to an insulated steel end ring). Whether random or layer wound, the coils nearly always have a six-sided diamond

[2]These coils are known as mush or random-wound coils because they are somewhat flexible and the positions of the individual turns are not precisely defined.

FIGURE 4-1   Magnetic circuit of a squirrel-cage rotor

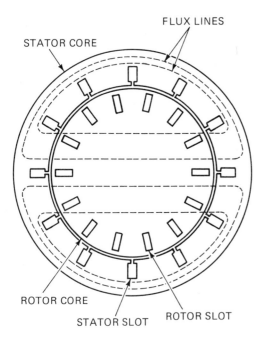

FIGURE 4-2   Open slot construction

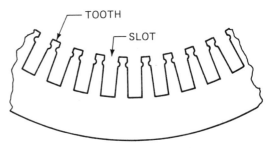

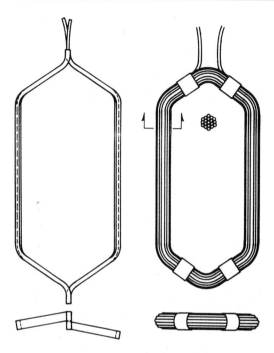

FIGURE 4-3   Preformed and mush coils before installation

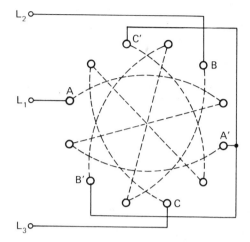

FIGURE 4-4   Elementary two-pole stator winding

shape, as shown in Figure 4-3, and may have several (from 3 to 40 or more) turns of wire.

The armature or rotor core is also built up from silicon steel laminations keyed to the shaft, and rotors can have any number of slots.[3] In contrast with the stator, the rotor slots are usually closed, as shown in Figure 4-1. On large machines a bare copper bar is inserted into each slot, and every bar is welded to two end rings, one at each end of the rotor core. The bars and end rings form what is known as a squirrel cage. On motors up to 400 kW or so, the squirrel cage will likely be a one-piece aluminum casting.

The rotor is supported on bearings so that it can rotate freely. The clearance between the rotor and stator cores is usually about 0.5 mm. Only the stator windings will be connected to the power supply. There are no external connections to the squirrel cage, and so there is no need for a commutator or slip rings, and no brushes.

The flux in a two-pole motor follows the figure-8 path shown in Figure 4-1. The flux paths in four- and six-pole motors are shown in Figure 4-4. The speed at which a three-phase motor runs is inversely proportional to the number of poles, which has to be an even number and seldom exceeds 14. However, there are few restrictions on the number of poles for which a given stator core can be wound or with which a given squirrel-cage rotor can be used. It is quite common to find four- six-, and eight-pole motors with identical squirrel cages and identical stator cores. The stator windings, however, are distinctly different, and we will examine them in more detail later in this chapter.

---

[3] We generally avoid having equal numbers of stator and rotor slots because, if their numbers are equal, the rotor and stator teeth tend to lock together magnetically, and this reduces starting torque. In most motors the rotor slots are skewed rather than the stator slots. Skewing either the rotor or stator slots reduces this locking tendency and also reduces magnetic noise when the motor is running.

Rotating Field in a Two-Pole Motor

In an elementary two-pole motor the windings for the three phases are spaced 120° mechanical from each other, as shown in Figure 4-4. We will name the three phases *A*, *B*, and *C*, and in the diagram the unprimed letters show the start of each phase and the primed letters indicate the finish of each phase. The small circles represent the conductors or coil sides in the slots. The curved dotted lines show the back connection between opposite sides of each coil, and the straight dotted lines represent the connections between the coils. For simplicity we have shown only two single-turn coils per phase (six coils in all). Multiple turns per coil and at least six coils per phase would be the usual case. We have shown the line connections (L_1, L_2, and L_3), and we have shown the phases as being wye connected, but a delta connection could also be used.

Looking at Figure 4-4, it is apparent that phase winding *A* can create an mmf only in the vertical direction, either up or down depending on the direction of current flow. If we specify directions in degrees measured counterclockwise from the right-hand horizontal, the mmf of phase *A* will be at either 90° or 270°. If we define positive current flow as into the motor at each line connection, positive mmf of phase *A* will be at 270°. Similarly, phase *B* creates an mmf in the 150° direction (−30° if the current is negative), and phase *C* works along the 210°–30° axis. The instantaneous mmf of any phase is proportional to the instantaneous current in that phase, and the phasor sum of their mmf's establishes the flux through the rotor (we will ignore rotor current at this point).

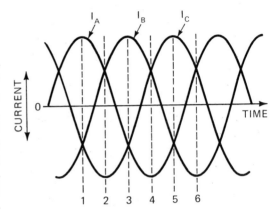

FIGURE 4-5   Three phase currents in a motor

Figure 4-5 shows the three phase currents that flow in the stator windings. When the current graph is above the zero line, the flow is into the motor; if below, the current is flowing out of the motor on that phase. If we pick the times marked 1, 2, 3, 4, 5, and 6 in Figure 4-5, and draw the necessary phasors to find the total mmf of the winding at each of those times, the result is as shown in Figure 4-6.

It is apparent that the three phase windings in the stator together create a constant-strength, smoothly rotating mmf, which in Figure 4-6 is traveling clockwise, and since the air gap is essentially uniform, they will establish constant-strength, smoothly rotating magnetic fields.

Two points need to be made here. First, if we reverse the phase sequence[4] of the stator currents so that they reach their positive peaks in the order *ACB* (rather than *ABC* as shown in Figure 4-5), the magnetic field will rotate the opposite way, and so the motor shaft rotation will reverse. Second, we

---

[4]The easiest way to reverse the phase sequence is to interchange any two of the three lines feeding the motor.

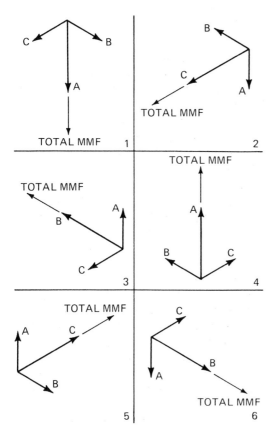

FIGURE 4-6 Rotating mmf of the stator windings

can see from Figure 4-6 that the magnetic field in our two-pole motor will turn one revolution for each cycle of our power supply. The speed at which the field rotates is known as the synchronous speed of the motor, and if we could change the frequency, the synchronous speed would vary accordingly. However, most power systems are of constant frequency, so we generally must regard synchronous speed as being constant for any given motor. For any number of poles, the synchronous speed is given by Equation 4-1. Some motors are designed so

that we can change the number of poles by changing the external connections, and this is the basis of two-, three-, and four-speed motors.

$$\text{speed (in rev/s)} = \frac{2f}{P} \qquad (4\text{-}1)$$

where f = frequency in hertz (Hz)
       P = number of poles

The rotating magnetic field does two things. First, it induces currents in the squirrel-cage bars so that torque will be developed on the rotor. Second, it induces a counter emf in the stator windings, and this action limits the current flow there. These two actions are essentially the same as those which occur in transformers, but let us consider some of their implications.

With all other factors constant, the emf (per turn) will be proportional to the amount of flux in the rotating field. Like a transformer, the emf in a motor is only moderately less than the applied voltage, and so the strength of the rotating field is proportional to the voltage per turn applied to the stator windings. If, for example, the stator windings are reconnected from series to two parallel paths and the applied voltage is cut in half, the rotating field remains the same and so does motor performance. This is the basis of dual-voltage motor windings.

If we reconnect the stator winding so that with the same applied voltage the voltage per turn is reduced (e.g., change from a delta to a wye connection), the rotating field becomes weaker and the motor torque will be reduced. This is the basis of triple-power-rated single-speed motors and also wye-delta starting.

It can be shown that if other factors remain constant, motor torque is proportional to the square of the voltage per turn. Because of this square relationship, induction motors are quite sensitive to voltage variations, and failure to start is often the first indication of a low-voltage condition.

If the applied frequency increases, the synchronous speed increases and tends to raise the cemf. However, the cemf cannot exceed the applied voltage, and so the flux decreases (varies inversely as the frequency). The decreased flux results in reduced motor torque. Motors rated at 50 Hz will theoretically run on 60 Hz, but unless the voltage is increased proportionally, the torque may be too low to drive the load.

Torque Development in a Squirrel-Cage Motor

Consider a squirrel-cage rotor in a rotating magnetic field as shown in Figure 4-7. If we apply the right-hand rule for induction, we find that the emf's in all the bars on the left-hand side of the armature are toward the observer, and the emf's in all the right-hand bars are away from the reader. Since all the bars are joined together at both ends, under normal operating conditions current flows in the squirrel-cage bars according to the crosses and dots in Figure 4-7. Since the

FIGURE 4-7   Voltage, current, and torque in a squirrel-cage rotor

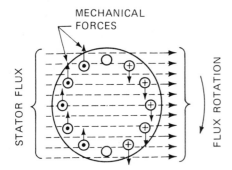

bars are carrying current and are surrounded by an external magnetic field, mechanical forces are generated on them as shown, and therefore torque is produced. This action is not greatly different from what happens in a dc motor.

The force on each bar is proportional to the product of bar current and surrounding flux density. However, in an induction motor, the voltage induced in the bar and therefore (with other factors constant) bar current is proportional to the voltage per turn applied to the stator windings. That is why the motor torque is proportional to the square of the voltage per turn.

It is apparent that we will have squirrel-cage current only if the rotating flux is cutting across the bars. If the squirrel cage were turning at synchronous speed, armature current and torque would be zero. For this reason, induction motors normally run below synchronous speed. The difference between synchronous speed and actual motor speed is called the slip and is often given as a fraction (or percentage) of synchronous speed. Under normal running conditions, slip is usually less than 10%, and with the rotor at a standstill, slip becomes 1.0 (or 100%). If the rotor is turning backward (relative to the rotating field), slip is greater than 1.0. It is important to understand what is meant by slip because so many rotor quantities are closely related to it.

Slip and Other Rotor Quantities

In the following discussion we will use the subscript A to indicate rotor quantities. E_A, I_A, f_A, Z_A, R_A, and X_A then represent rotor voltage, rotor current, rotor frequency, impedance, resistance, and reactance, respectively. Except for R_A, these quantities are variable, depending on the value of slip. It is often convenient to designate these values when the slip is unity. To do so, we will use the subscripts *BA*. Just as a memory device, think of the subscript *A* as originating from the words "armature" and "actual" rather than reflected values, which are dealt with later. The subscript *B* means "blocked"; that is, the rotor is unable to turn. We will use the letter *s* to indicate slip as a fraction of synchronous speed. We will use the subscript *s* to indicate stator quantities.

Slip and Rotor Voltage

At zero slip, the rotor voltage must be zero. At 100% slip, the voltage per bar in the rotor (E_{BA}) becomes equal to the cemf per conductor in the stator (both voltages are generated by the same flux). If the rotor voltage is known at any value of slip, we can find E_A at any other value of slip using Equation 4-2.

$$|E_A| = s|E_{BA}| \qquad (4-2)$$

Slip and Rotor Frequency

If we consider any one rotor bar, the greater the slip, the more rapidly the north and south magnetic poles of the stator are going by and, therefore, the higher the frequency of the rotor voltage. If we look at the stator for a moment, the frequency of the cemf must be the same as the applied voltage. Since cemf and E_{BA} are produced by the same flux, f_{BA} must be equal to the frequency applied to the stator windings. We therefore have Equation 4-3.

$$f_A = sf_s \qquad (4-3)$$

Slip and Rotor Reactance

Current flow in a squirrel-cage bar tends to set up flux (called leakage flux) around it,

and so each bar has a fixed amount of inductance (or leakage inductance). The manufacturer can control the inductance of the squirrel cage by changing the cross-sectional shape of the bars and/or raise the inductance of the bars by burying them at a greater depth in the rotor iron (providing a better path for leakage flux). It is convenient, however, to regard the inductance of a given rotor as a constant.

The value of X_A, however, is another story. Because inductive reactance is proportional to frequency and f_A is proportional to slip, we can write Equation 4-4.

$$X_A = sX_{BA} \qquad (4-4)$$

Slip and Rotor Impedance

The standard equation for impedance applies to rotors as shown by Equation 4-5.

$$|Z_A| = \sqrt{R_A^2 + X_A^2} \qquad (4-5)$$

At low values of slip, $X_A \ll R_A$, and so Z_A tends to be constant. At high values of slip, $X_A \gg R_A$, and $|Z_A|$ becomes almost directly proportional to the slip.

Slip and Rotor Current

I_A always equals E_A/Z_A, but at low values of slip $|Z_A|$ is almost constant, so $|I_A|$ becomes almost directly proportional to slip. At high values of slip, $|E_A|$ is proportional to slip and $|Z_A|$ is almost so, and $|I_A|$ then tends to be constant. Figure 4-8 shows what to expect. Rotor inductance (which creates X_A) establishes the maximum value of $|I_A|$.

Slip and Rotor Power Factor

At low values, $X_A \ll R_A$ and so I_A is practically in phase with E_A. As the slip increases, X_A becomes significant and $|I_A|$ lags $|E_A|$, and this phase angle can approach the theoretical limit of 90°. Rotor power

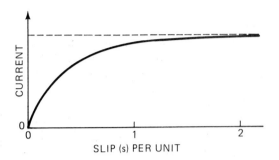

FIGURE 4-8   Rotor current at various values of slip

factor is related to this angle as shown by Equation 4–6.

$$\text{Rotor power factor} = \cos \angle \frac{|E_A|}{|I_A|}$$

$$= \frac{R_A}{|Z_A|} \qquad (4\text{–}6)$$

Figure 4-9 shows the loci of $|E_A|$ and $|I_A|$ as slip increases.

Rotor Power Factor and Motor Torque

Figure 4-7 is realistic for a squirrel cage where $X_A \ll R_A$; but if X_A is substantial, the currents lag the voltages in the rotor bars as shown in Figure 4-10. If we apply the left-hand rule to obtain the force directions, it is apparent that the force on some of the bars will oppose the torque produced by the majority, and that is why the net torque is proportional to the rotor power factor.

The relationship between torque, field flux, rotor current, and rotor power factor is given by Equation 4–7, which is known as the fundamental torque equation for an induction motor.

$$T = K\phi |I_A| \cos \theta_A \qquad (4\text{–}7)$$

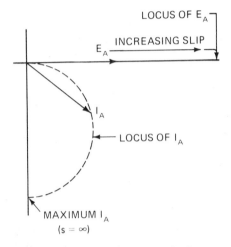

FIGURE 4-9   Loci of E_A and I_A as slip changes

where $\theta_A = \angle \dfrac{|E_A|}{|I_A|}$

K = all other factors that are constant for a given motor

At low values of slip, rotor power factor is near unity and $|I_A|$ is very nearly proportional to slip. Since ϕ is constant, torque increases with slip. However, at high values of slip, $|I_A|$ levels off and $\cos \theta_A$ decreases drastically, causing a decrease of torque. An elementary squirrel cage (with constant resistance and inductance) therefore al-

FIGURE 4-10   Rotor currents and forces when I_A lags E_A

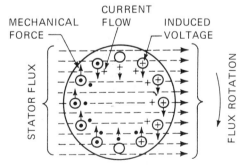

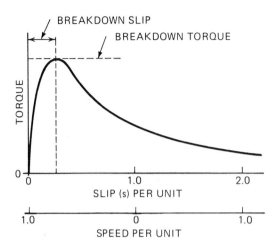

FIGURE 4-11 Torque-slip curve for an elementary squirrel cage

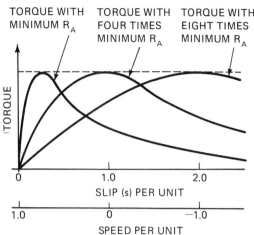

FIGURE 4-12 Effect of rotor resistance on the torque curve

ways has a torque curve shaped like that in Figure 4-11.

The highest point on the torque curve is known as the breakdown torque of the motor and usually occurs at or above 80% of synchronous speed. Neglecting the effect of resistance and reactance in the stator windings, breakdown torque occurs when $X_A = R_A$. Reducing the rotor inductance will raise the breakdown torque and also the value of slip at which it occurs.

Rotor Resistance and Motor Torque

If we increase rotor resistance and slip in the same proportion, rotor voltage and rotor impedance will both change in the same proportion. We therefore end up with the same rotor current at the same power factor, and so the motor torque is unchanged.

To put this another way, for a given value of torque, slip is proportional to rotor resistance. Adding rotor resistance will not lower the torque curve but merely stretches

it out so that the same torque values occur at lower speeds, as shown in Figure 4-12.

Effects of Rotor Currents

The mmf of the squirrel-cage currents tends to modify the revolving field that is created by the stator windings. However, any change of the rotating field will change the cemf, and this permits a change of stator current that practically offsets the mmf of the rotor. The result is that (if we neglect the resistance and leakage reactance of the stator winding) the rotating field is unchanged by the rotor mmf, and the rotor currents are directly reflected as increased current in the stator circuit.

Let us define stator current (I_s) as the total current that the motor draws from the line. Similarly, we will define I_o as the exciting current that sets up the rotating magnetic field and supplies the iron losses, and I_R will represent reflected rotor current. The relationship between I_s, I_o, and I_R is given by Equation 4–8 and shown by the

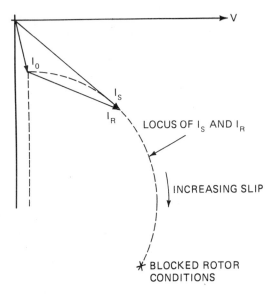

FIGURE 4-13   Effect of slip on stator current

phasor diagram in Figure 4-13. Compare Figures 4-13 and 4-9. We can now see why

$$I_s = I_o + I_R \qquad (4-8)$$

increasing the load on the motor causes it to draw more line current. We can understand why the starting current will be higher than normal running current and that it is controlled mainly by the line frequency impedance of the squirrel cage. We can also see that the no-load power factor is poor, but it improves with load and deteriorates again with gross overload.

Figure 4-13 is a simple version of the circle diagram[5] for an induction motor, and, if elaborated, it gives the same results as the L-shaped equivalent circuit that is treated later in this chapter. In practice, the pres-

[5]See section 18-56 of Donald G. Fink and John M. Corrall, *Standard Handbook for Electrical Engineers*, 10th ed. (New York: McGraw-Hill Book Company, 1968).

ence of leakage reactance and resistance in the stator circuit causes I_o to decrease under load, resulting in a decrease of field flux. The Steinmetz equivalent circuit (treated later) takes this factor into account.

Current-Displacement Effect

Consider a squirrel-cage bar of deep, narrow cross section as shown in Figure 4-14. If current flows in the bar, leakage flux will be set up around it approximately as shown. Because the inner side of the bar is looped by the most flux, that side of the bar will have the most inductance. At low values of slip this is not of much significance, but at high slip the inner side of the bar has more inductive reactance, which tends to choke out current on that side of the bar. This action is essentially the well-known skin effect, but we have given it the name "current

FIGURE 4-14   Leakage flux around a deep, narrow rotor bar

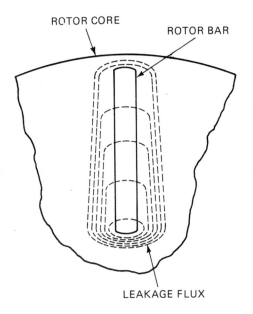

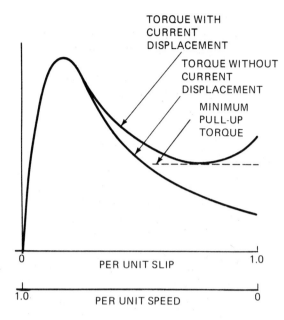

FIGURE 4-15   Effect of current displacement on motor torque

displacement" because the current flows mainly on the outside edge of the bar, nearest the rotor surface. This action raises the effective value of R_A, which simultaneously reduces starting current and improves starting torque, as shown in Figure 4-15.

Note that if the current-displacement effect is substantial, the torque curve is shown as having a noticeable sag at about one quarter of synchronous speed. This is a fairly common observation, and the low point on the torque curve is called the minimum pull-up torque.

Double Squirrel-Cage Rotor

The inductance of a squirrel cage is determined primarily by the depth at which it is buried in the rotor iron. The resistance is determined by the cross-sectional area and resistivity of the material used for the bars and end rings.

It is quite practical to place a double or even what amounts to a triple squirrel cage on a single rotor, as shown in Figure 4-16. The squirrel cages will probably share the same end rings. The outer cage has more resistance than the inner cage, but less inductance. At high slip, most of the current flows in the high-resistance cage, providing good starting torque. At low slip most of the current will be in the low-resistance bars so that efficiency will be high and full-load slip will be low. This shift of current from the outer to the inner bars is essentially the same as the current-displacement effect, but double or triple cages provide a better torque curve than deep, narrow bars.

Double and triple squirrel cages have one deficiency. Because both cages must be made of the same material, the only way to get high resistance in the outer cage is to use bars of small cross section. Because of their small volume of metal and because they carry most of the current, the outer bars heat up rapidly during the starting period. For this reason, double and triple squirrel cages are not the best design for frequent starting, particularly with high-inertia loads.

FIGURE 4-16   Double and triple squirrel cages

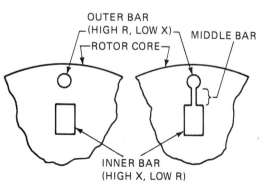

Operation Above Synchronous Speed

If a squirrel-cage motor is driven above synchronous speed, the relative motion between the stator flux and the rotor bars is reversed, so the squirrel cage currents are reversed. As a result, two things happen:

Generator Action

Since the rotor currents are reversed, the currents reflected into the stator are reversed, and so the motor delivers power (kilowatts) back to the line. When used this way, the machine is known as an induction generator. Its most serious limitation is the fact that it continues to draw a magnetizing current from the line, and so it cannot be used alone but only in parallel with synchronous alternators. Historically, the induction generator has seen little use. However, they are presently offered as part of some small, packaged hydroturbine and generator sets.

Braking Action

The reversed rotor currents cause the motor to develop braking torque. This action is known as regenerative braking. From the induction generator standpoint this is just the law of energy conservation, but it is also important from a "motor and load" point of view. If the load tends to drive the motor at times, the motor will not permit much increase of speed. The torque curve of an induction motor is symmetrical about the synchronous speed point, as shown in Figure 4-17.

Wound-Rotor Motor

Wound-rotor motors use the same stator windings as squirrel-cage motors, but instead of a squirrel cage, the rotor has an in-

sulated three-phase winding connected to three slip rings that are mounted on the shaft. The machine has brushes that ride on the slip rings, and if the brushes are short circuited together, the motor runs essentially like a squirrel-cage type. But if we connect resistors between the slip rings, this has the effect of increasing R_A, which stretches the torque curve toward increased slip. If we use the correct amount of resistance (so that $R_A = X_{BA}$), breakdown torque will occur at a standstill, so that maximum possible starting torque will be obtained. If we use some kind of three-phase rheostat in the rotor circuit and cut out resistance in the appropriate manner as the motor accelerates, the motor will develop breakdown torque all the way up to about 80% of synchronous speed. Adding rotor circuit resistance reduces starting current, but too much resistance reduces starting torque, which may not be acceptable. With a given torque load, the operating speed will decrease as rotor resistance increases.

FIGURE 4-17 Typical torque curve of an induction motor

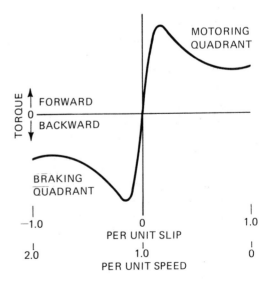

The wound rotor has three main advantages over a squirrel-cage type. First, it will develop high starting and accelerating torque. During acceleration, most of the heating occurs in the control resistors where it can be more readily dissipated, and so the wound-rotor machine is well adapted to frequent starting of high-inertia loads. Second, with the rotor resistance set to get maximum starting torque, it draws less starting current than a squirrel-cage motor. Third, it offers the possibility of virtually stepless speed control.

Wound-rotor motors have three disadvantages. First, the efficiency is almost directly proportional to the operating speed. At 50% of synchronous speed, efficiency is virtually cut in half, and this becomes important if the motor runs at low speed for long periods. Second, the controller does not change the no-load speed very much, and so at low speed settings the speed regulation becomes very poor. Third, they cost much more than squirrel-cage motors.

Because of its construction, a wound-rotor machine is sometimes known as a slip-ring motor. Because of the similarity between induction motors and transformers, the rotor circuit (be it squirrel cage or wound) is sometimes called the secondary circuit.

Losses and Efficiency of Induction Motors

The standard definition of efficiency (output/ input) applies to induction motors and so does the relationship between torque, power, and speed given by Equation 2–3. The losses in an induction motor and the relationship between some of the more significant quantities are shown in Figure 4-18. Of all the power supplied to the stator windings (P_{in}), some of it is dissipated as copper loss in

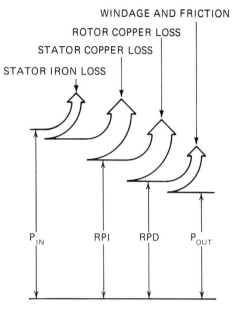

FIGURE 4-18   Power-flow diagram for an induction motor

those windings and some of it is expended as stator iron losses (hysteresis and eddy currents in the stator core). The balance is transferred to the rotor via the rotating magnetic field and is known as the rotor power input (abbreviated RPI). The relative motion between the rotating flux and the rotor core is very slow under normal running conditions, and so the iron loss in the rotor core is negligible.

It is shown later in this chapter that the rotor copper loss is always equal to the product of RPI and slip. If we subtract rotor copper loss from RPI, we get the rotor power developed (abbreviated RPD), and if we subtract windage and friction losses from RPD, we have the power output of the motor (P_{out}). For slip-ring motors, the rotor copper loss in Figure 4-18 includes the losses in the secondary resistors.

SOME ASPECTS OF STATOR WINDING DESIGN

Conventional Stator Windings

The stator winding shown in Figure 4-4 differs from the usual arrangement in two respects. First, there are only half as many coils as there are slots in the stator core, and each slot contains only one side of one coil. This is a "basket winding" and is not very popular in North America. The usual (conventional) arrangement is to have as many coils as there are slots, with each slot containing the left-hand side of one coil and the right-hand side of some other coil.[6] The second difference is the larger number of slots and coils that are usually employed. Figure 4-19 shows how the coils are installed in a conventional stator. Ideally, the mmf of each phase should be sinusoidally distributed around the air gap, and the larger number of coils enables us to more closely approach this condition.

No matter how many magnetic poles the stator winding will create, the coils are all basically installed as shown in Figure 4-19. The only significant difference is the pitch or span of each coil, that is, the distance between the opposite sides of each coil. The coil span in Figure 4-19 is six slots.[7] As we did on dc armature windings, we will define the angular displacement between the center lines of adjacent poles to be 180 electrical space degrees. We use the word electrical to prevent confusion with mechanical space

degrees, and we include the word "space" here to distinguish this angular measurement from electrical time degrees, which were used in Chapter 3 and have at least been implied in this chapter. This distinction between electrical space and electrical time is not always maintained either in this book or in casual conversation. It is often simply written (or said) "electrical degrees," and the reader (or listener) must determine from the context whether the reference is to space or time. Words like "chorded," "full pitch," or "fractional pitch" have the same meaning here as they had in Chapter 2. Coil spans greater than 180 electrical space degrees have exactly the same magnetic effect as coil spans of less than 180°, but the lat-

FIGURE 4-19   Partially wound conventional stator

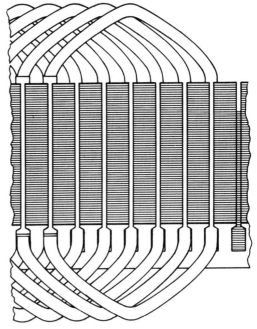

[6]Because of the way the coils are installed, one side of each coil lies in the bottom half of one slot and the top half of some other slot. To indicate one side of a coil, we therefore often refer to the "top" or "bottom" side, respectively. The top and bottom coils in any one slot may belong to different phases and have considerable voltage between them. They are therefore usually separated by additional pieces of solid insulaton.

[7]This coil span could be specified as "1 and 7."

ter is preferable because it shortens the ends of the coils, reduces the amount of wire used, and cuts down the winding resistance. About five-sixths of full pitch is usual.

To obtain a rotating field that has (*P*) poles, each phase of the stator winding must create (*P*) poles. We therefore divide the coils into (*3P*) groups of coils that are known as pole-phase groups or phase belts, and wire them as required. For a 4-pole motor with 36 coils, we divide them into 12 groups of 3 coils each. For a 2-pole motor with 36 coils, we have 6 pole-phase groups of 6 coils each. The coils in each pole-phase group are wired in series, with their mmf's all in the same direction (i.e., aiding each other). If we number the pole-phase groups consecutively around the stator bore, groups 1, 4, 7, 10, 13, etc., will belong to one phase, groups 2, 5, 8, 11, 14, etc., will belong to another phase, and groups 3, 6, 9, 12, 15, etc., will belong to the remaining phase. However, a pole-phase group occupies only 60° of space in a conventional winding, and the three windings must always be 120° electrical from each other. It is therefore convenient to think of one phase as starting on coil group 1 and continuing through the others of that series, the second phase starting on coil group 3 and continuing through all the other members of that winding, and the third phase as starting on group 5 and ending on group 2. Done this way, all three windings are symmetrical. It is possible to start each phase winding on any coil group belonging to that phase, but this may result in a loss of symmetry and leads to confusion.

The pole-phase groups may be connected in series or in a number of parallel paths and in either a wye or a delta. From a repair-shop point of view, changing the number of parallel paths or changing from wye to delta will change the voltage at which the motor should operate. From a design point of view, these changes permit changes in the number of turns and size of wire used for the coils and may permit a more economical design with no change of motor performance.

Let us define the positive direction of current flow to be from start to finish in each phase (i.e., line to wye point if wye connected; the same way around the delta if delta connected). No matter whether the winding is wye or delta, or how many parallel paths it has, with positive current in all three phases, adjacent pole-phase groups must have opposite magnetic polarity.

Stator Winding Diagrams

There are two ways to draw wiring diagrams that show how the stator windings are connected. One way is to draw the stator winding as if it were laid out on a flat surface. Figure 4-20 shows a stator with 24 coils connected into 6 groups of 4 coils each. The coils appear to have only one turn, but they may have many. The coil span is 11, and the reader must note that this can be seen by looking at the back end of the coils (the top side of the diagram, away from the coil or group leads). The top side of each coil is shown as a solid line, and the bottom side of each coil is shown as a dotted line. The interconnections between the pole-phase groups can easily be added as desired.

In the other type of stator winding diagram the pole-phase groups are represented by segments of a circle, as shown in Figure 4-21. The ends of the segments represent the start and finish of each pole-phase group, and we can arbitrarily assume that current flow clockwise through a segment means that the pole-phase group creates a

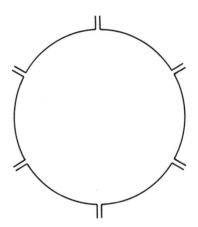

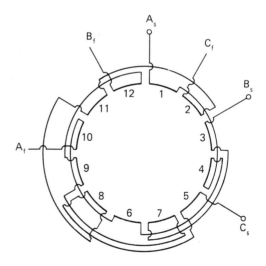

FIGURE 4-20   Stator coils connected into pole-phase groups

north magnetic pole. This is the type of wiring diagram that we will use in this book.

When less detailed information will suffice, we often use schematic diagrams of motor windings such as those in Figure 4-32.

Series-Connected Stator Windings

Let us give the three phases of the stator winding the names A, B, and C, and use these letters with the subscripts s and f to

identify the start and finish of each winding. Normally, these motors have only three external leads.

In most series circuits, the order in which the parts are connected does not matter, but in motor windings we always follow one of two patterns. Figure 4-22 shows a "1 to 4" connection. The circuit for phase A goes through group 1 forward, then back-

FIGURE 4-22   Series stator winding with a "1 to 4" connection

FIGURE 4-21   Pole-phase groups in two- and six-pole stators

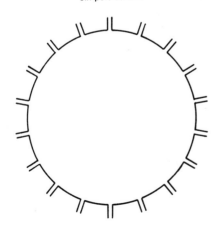

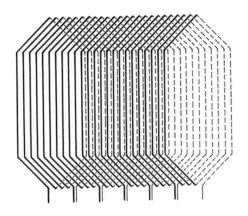

ward through group 4, forward through group 7, etc., until all the groups belonging to that phase have been connected into one series path. The other two phases are connected in identical fashion, starting on groups 3 and 5, respectively. The windings are shown in a wye (the dotted connection) but could just as easily be connected delta.

Figure 4-23 shows a "1 to 7" connection. The circuit for phase *A* goes through group 1 forward, then group 7 forward, then backward through groups 10 and 4. Again, phases *B* and *C* are identical, starting on groups 3 and 5. We have shown the jumpers necessary to make a delta connection as dotted lines, but a wye connection is equally possible. The "1 to 7" connection is mandatory for two-speed consequent pole motor windings.

Rotating Field in a Four-Pole Motor

Figure 4-24 shows the currents in the three phases of the stator windings in Figure 4-22 or 4-23. At time 1 in Figure 4-24, we can trace the direction of current in each phase and show that coil groups 1, 2, and 3 are creating a north magnetic pole, and so are groups 7, 8, and 9. Coil groups 4, 5, and 6 and groups 10, 11, and 12 will be setting up south poles in their areas. A four-pole magnetic field is therefore created. At time 2, groups 2, 3, and 4 and 8, 9, and 10 will be setting up north poles, while groups 11, 12, and 13 and 5, 6, and 7 will be creating south poles, so the magnetic field has rotated clockwise 60 electrical space degrees. At time 3, north poles are established by groups 3, 4, and 5 and 9, 10, and 11, while the two remaining clusters of three groups each set up south poles. We can see that the field rotation in electrical space degrees will

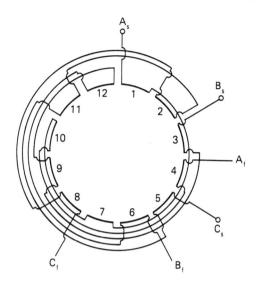

FIGURE 4-23 Series stator with a "1 to 7" connection

always equal the elapsed time in electrical time degrees.

Parallel Connections and Nonuniform Pole-Phase Groups

If a winding is to be connected in two or more parallel paths, there is one basic rule

FIGURE 4-24 Currents in the stator windings of a motor

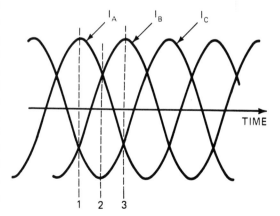

to follow. Be sure that the cemf in each path will be the same. Otherwise, curents will circulate between the paths, causing excessive heating. When all the pole-phase groups are the same, this is a matter of getting the same number of groups in series in each path. Since there is no way to split a pole-phase group and get equal cemf's in each path (the cemf's always differ either in magnitude or phase position in this case), the number of parallel paths must be an integral factor of the number of poles. An eight-pole motor winding can therefore usually be connected in one, two, four, or eight parallel paths, but not in three, five, or six, and if the pole-phase groups are not uniform, additional restrictions on the possible number of parallel paths may arise.

Motor manufacturers find it economically desirable to use the same stator cores (or at least the same stator laminations or punchings) to build a wide variety of motors, and this sometimes results in nonuniform pole-phase groups. Satisfactory operation can be obtained, but we must end up with equal numbers of coils in each phase, and the two sizes of pole-phase groups must be properly distributed around the stator bore.

If nonuniform pole-phase groups are required, there will be only two sizes of groups and only one coil difference between a large group and a small one. Sometimes the majority of the groups will be large, and sometimes the large groups will be the minority; but if we let n_1 equal the minority number of groups and n_2 equal the majority number of groups, then $n_1 + n_2$ will equal total number of coil groups in the motor. In general, if $(n_1 + n_2)/n_1$ is any integral multiple of 3, we must assign equal numbers of coils per pole and then rotate the minority group uniformly among the phases. If $(n_1 + n_2)/n_1$ is not a multiple of 3, then the coil groups can

be arranged symmetrically around the stator bore. Consider a stator core that is to be wound with 36 coils. Four- and six-pole windings work out to 3 and 2 coils per group, respectively, but for an eight-pole winding we require 24 groups, so we end up with 12 groups of two coils each and 12 groups of one coil each. Since $24/12 = 2$, we can spread these uniformly around the stator, and the number of coils per group ends up to be 1, 2, 1, 2, etc. If we check back, we find that doing this leaves equal numbers of coils in each phase (four 1's and four 2's), so that is the way to go. However, this winding cannot be hooked up in eight parallel paths because we cannot make the paths equal (one, two, or four parallel paths could be done).

If we try to use 36 coils for a 10-pole winding, we must have a total of 30 coil groups, so we end up with 24 groups of 1 and 6 groups of 2. Since $30/6 = 5$, we end up with groups 5, 10, 15, 20, 25, and 30 having 2 coils each and all the others having 1. Again this distributes the coils equally among the phases, so it is the best possible arrangement. However, each phase contains two groups of two and eight groups of one coil, and we cannot get more than two equal parallel paths in such a case.

If we wish to use 48 coils for a 6-pole winding, we end up with 12 groups of 3 coils and 6 groups of 2 coils. But note that $18/6 = 3$, and if we tried to distribute these coils symmetrically around the stator we would end up with 3, 3, 2, 3, 3, 2, etc., and this would leave one phase with fewer coils than the other two. So we must divide the pole-phase groups into clusters of three and assign equal numbers of coils (in this case 8) to each cluster. We now can assign the correct number of coils to each group, being careful to rotate the odd group uniformly among the phases. The grouping then ends up to be 3, 3, 2, 3, 2, 3, 2, 3, 3, 3, 3, 2, 3, 2, 3, 2, 3, 3, as shown in Figure 4-25. Since each phase contains four groups of three and two groups of two coils each, it is not possible to connect this winding into more than two parallel paths.

Dual-Voltage Motors

If a motor is reconnected to get twice as many parallel paths in the stator winding, it will perform exactly the same as it did before the change provided that we apply only half as much voltage. However, the line current at the lower voltage will be double that which was formerly expected.

A dual-voltage motor is one that has the necessary external leads provided so that it can be easily connected for operation at either of two voltages without opening the motor. Because both voltages are in common use, manufacturers commonly build motors that can be connected for use at either 460 or 230 V.[8] This minimizes the number of different motors that must be manufactured and stocked. The National Electrical Manufacturer's Association (NEMA) and the Electrical and Electronic Manufacturer's Association of Canada (EEMAC)[9] have adopted two standard ways of identifying the leads on dual-voltage motors (one for a wye and one for a delta) using the numbers 1 to 9 inclusive.

To change a 460-V series-connected motor

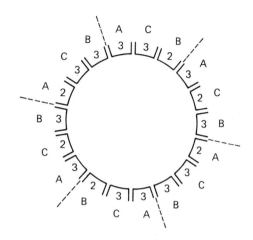

FIGURE 4-25   Distribution of 48 coils into 18 pole-phase groups

to a dual-voltage connection and end up with standard lead numbers, proceed as follows. Identify the start of phase *A* as lead 1. Trace the winding halfway through, cut the circuit at that point, and bring the cut ends out as external leads numbered 4 and 7, where 4 is the lead that has continuity to 1. Label the start of phase *B* as lead 2, cut that phase in the middle, and label the cut ends 5 and 8, with 5 having continuity to 2. The start of phase *C* is then labeled 3, and the cut ends in the center of that phase become 6 and 9. This basic procedure works for either a 1 to 4 or a 1 to 7 connection and either a wye or a delta. If we follow it on Figures 4-22 and 4-23, we end up with the winding shown in Figures 4-26 and 4-28. Figures 4-27 and 4-29 show how to connect these motors for each voltage.

Triple-Power-Rated Motors

The stator windings of triple-power-rated motors are usually designed so that they can be operated in delta, and if connected that way, the motor develops its highest torque

---

[8]We generally think of this as changing from series to two parallel paths, but actually the change may involve going from two to four parallel paths or from three to six. However, this makes no difference to the external wiring.

[9]Most of the major motor manufacturers in Canada are subsidiaries of parent firms in the United States. For this reason, industry-sponsored standards in Canada are very similar (although not always identical) to those found in the United States.

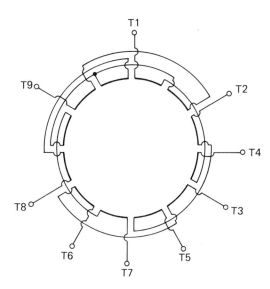

FIGURE 4-26   Dual-voltage wye-connected motor

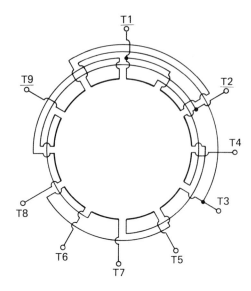

FIGURE 4-28   Dual-voltage delta-connected motor

FIGURE 4-27   Connections for a dual-voltage
wye-connected motor

FIGURE 4-29   Connections for a dual-voltage
delta-connected motor

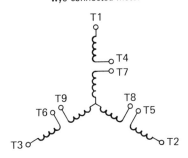

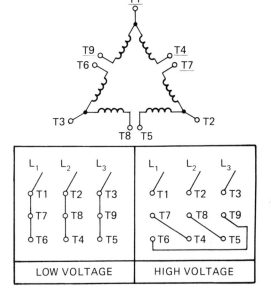

and power. However, the windings are provided with taps so that they can be reconnected in a way that reduces the voltage per turn. This lowers the input current and the torque curve so that the motor performs like one of the same speed but of a lower power rating.

Figure 4-30 is an example of a triple-power-rated motor winding. In this example the taps are located at the one-quarter and three-quarter points in the winding, but this is not always the case. The figure shows how this motor should be connected for each of its power ratings.

A triple-power-rated motor may be desirable if (1) the utility company charges a flat rate for energy based upon X \$ per kW output per month, *and* (2) the motor must be fitted to the driven machine beforehand, but the power required to drive the machine will not be known until the field installation is complete. This is why oil-well-pump motors are often triple power rated.

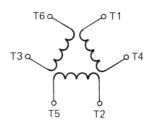

FIGURE 4-31 Wye-start delta-run motor

Motors for Wye-Delta Starting

The stator windings of these motors are designed to work delta connected, but to reduce starting current, the motor is connected in a wye during the starting period. To make the change from wye to delta, the start and finish of each phase must be brought out as an external lead; otherwise, these motors are not different from those shown in Figures 4-22 and 4-23. Figure 4-31 shows the standard numbering of the external leads. The "wye start–delta run" scheme reduces starting current and starting torque to one third of the normal values.

FIGURE 4-30 A triple-power-rated motor

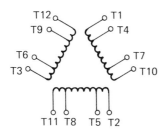

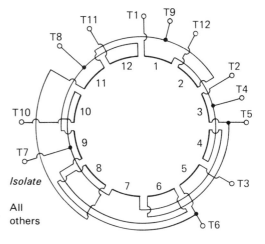

Power Output	Line Connections	Join and Isolate	Isolate
Maximum	L_1–1, L_2–2, L_3–3	10–2, 11–3, 12–1	
Medium	L_1–1, L_2–2, L_3–3	10–5, 11–6, 12–4	All others
Minimum	L_1–1, L_2–2, L_3–3	10–8, 11–9, 12–7	

Motors for Part Winding Starting

The windings in these motors are connected in two parallel paths, and to reduce the starting current, only one path is energized during the starting period. The motor is usually connected in the 1 to 7 pattern. If wye connection is used, the schematic and wiring diagram and the numbering of the external leads are identical to a dual-voltage wye-connected motor. However, sometimes leads 4, 5, and 6 will be joined inside the motor so that they are not accessible, and the winding becomes two isolated wyes. If delta connected, the windings will form two isolated deltas, as shown in Figure 4-32.

Consequent-Pole Windings

The stator core and coils are essentially the same as conventional windings, but the coils are divided into only half as many pole-phase groups. To get (P) poles in the rotating field, we use only $3P/2$ coil groups, and so this is sometimes known as a half-coiled winding. Each pole-phase group will now occupy 120 electrical space degrees.

All the coil groups in each phase are connected to obtain the same magnetic polarity. It is therefore possible to start the phase windings at any convenient coil group, and in Figure 4-33 we have started on groups 1, 2, and 3. A wye connection has been shown, but a delta connection is equally possible.

Consequent-pole windings are almost never used for single-speed motors, but many two-speed motors operate as consequent-pole machines on their lower speed, and so the operation must be understood.

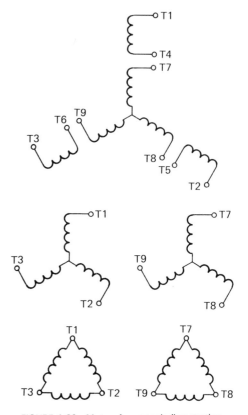

FIGURE 4-32   Motors for part winding starting
FIGURE 4-33   Consequent-pole motor winding
(eight poles)

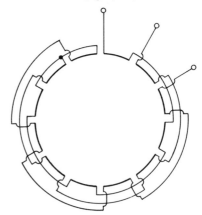

Operation of Consequent-Pole Windings

By referring to Figures 4-24 and 4-33, we can see how a rotating magnetic field will be produced by a consequent-pole winding. At time 1, coil groups 12 and 1, 3 and 4, 6 and 7, and 9 and 10 will create north poles, and groups 2, 5, 8, and 11 will create south poles. At time 2, we have groups 1, 4, 7, and 10 creating north poles and groups 2 and 3, 5 and 6, 8 and 9, and 11 and 12 forming south poles. At time 3, north poles are formed by groups 1 and 2, 4 and 5, 7 and 8, and 10 and 11, while south poles appear at groups 3, 6, 9, and 12. We can now see that the field has eight poles and rotates clockwise.

Changing from Conventional to Consequent Pole

Figure 4-34 shows two ways in which a winding can be changed from conventional to consequent-pole operation. One way is to wire the coil groups so that current flow from start to finish provides consequent-pole operation. Current flow from the center tap toward both ends will then provide conventional operation. This is done on pole-changing types of constant-torque and variable-torque motors.

The other possibility is to wire the coil groups so that current flow from start to finish provides conventional operation, and current flow from the center tap toward both ends provides consequent-pole operation. This is done on consequent pole constant-power motors. It should be noted that no matter which of these pole-changing circuits is used the 1 to 7 connection is essential.

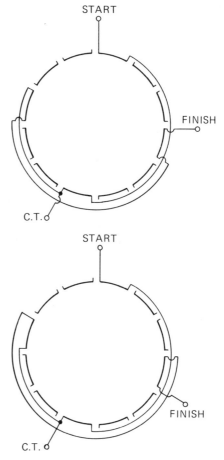

FIGURE 4-34   Two-pole changing windings (one-phase only)

Multispeed Motors

A multispeed motor is one that can operate at any one of two, three, or four definite speeds. The practical limit to the number of speeds is four, and there are three types of these motors.

A constant-power motor has the same rated power output at any of its rated speeds.

A constant-torque motor has the same full-load torque at any of its rated speeds, and so the power ratings become proportional to the synchronous speeds. The full-load torque of a variable-torque motor is proportional to its synchronous speeds, and its power ratings are therefore proportional to the square of the synchronous speeds.

Whether constant-power, constant-torque, or variable-torque characteristics are desired, the speed change can be obtained in either or both of two ways.

Use of Two Stator Windings

If a motor is provided with two stator windings that are wound for different numbers of poles, two-speed operation can be obtained by energizing the windings separately. It is not practical to use more than two windings.

Whenever two stator windings are installed, it is almost inevitable that when the motor is running on one winding the rotating field will induce voltages in the other winding. If the unused winding forms closed loops (e.g., it has parallel paths or is a closed delta connection), current will circulate in that winding, causing a loss of power and rapid overheating. For this reason, the windings must be connected either in a series wye or else a broken delta connection, as shown in Figure 4-35. The wye is preferred because the delta requires extra contacts in the control equipment to break each delta when it is not in use. Two-speed motors will have two windings only if the ratio of the synchronous speeds is not 2 to 1.

Use of Pole-Changing Windings

Two synchronous speeds with a 2 to 1 ratio can be readily obtained with one pole-

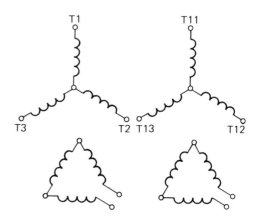

FIGURE 4-35   Two-speed two winding motors

changing winding. These are known as two-speed consequent-pole motors. If three speeds are required, one pole-changing winding and one single-speed winding will be used; for four speeds, two pole-changing windings are required.

There are three types of pole-changing motors:

CONSTANT-POWER MOTORS These run with the windings delta connected for high speed and two parallel wye connected for low speed. A wiring diagram for a four- to eight-pole motor, a schematic diagram, and connecting instructions are given in Figure 4-36. In three- and four-speed motors the delta is broken beside lead 5, and lead 7 is added.

CONSTANT-TORQUE MOTORS These run with the windings connected two parallel wye for the high speed and series delta for the low speed. A wiring diagram for a four- to eight-pole motor, a schematic diagram, and connecting instructions are given in Figure 4-37. In three- and four-speed motors, the closed loop is broken beside lead 3, and lead 7 is added.

is easiest to draw wiring diagrams of pole-changing windings if we think of them in their series configuration. A constant-power motor on its series connection is just like a conventional delta-connected winding with center taps. A constant-torque motor becomes a consequent-pole delta-connected motor with center taps on each phase, and a variable-torque motor is just a consequent-pole wye-connected motor with center taps.

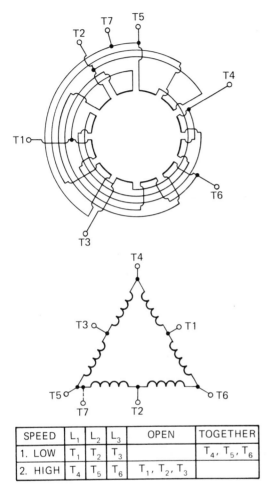

SPEED	L$_1$	L$_2$	L$_3$	OPEN	TOGETHER
1. LOW	T$_1$	T$_2$	T$_3$		T$_4$, T$_5$, T$_6$
2. HIGH	T$_4$	T$_5$	T$_6$	T$_1$, T$_2$, T$_3$	

FIGURE 4-36   Constant-power motor

FIGURE 4-37   Constant-torque motor

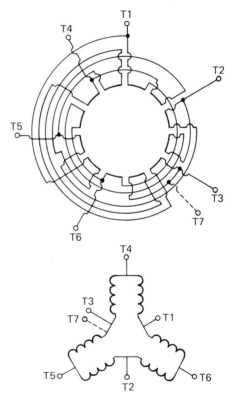

SPEED	L$_1$	L$_2$	L$_3$	OPEN	TOGETHER
1. LOW	T$_1$	T$_2$	T$_3$	T$_4$, T$_5$, T$_6$	
2. HIGH	T$_6$	T$_4$	T$_5$		T$_1$, T$_2$, T$_3$

VARIABLE-TORQUE MOTORS   These motors run with the windings connected two parallel wye for the high speed and series wye for the low speed. A wiring diagram for a four- to eight-pole motor, a schematic diagram, and connecting instructions are given in Figure 4-38.

All pole-changing windings use a 1 to 7 connection, and we always start the phase windings on coil groups 1, 3, and 5. But it

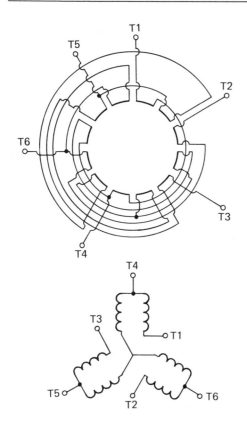

SPEED	L_1	L_2	L_3	OPEN	TOGETHER
1. LOW	T_1	T_2	T_3	T_4, T_5, T_6	
2. HIGH	T_6	T_4	T_5		T_1, T_2, T_3

The choice of coil span is more critical for a pole-changing winding than it is for an ordinary type. A coil span equal to full pitch at high speed is impractical because it results in a coil span of twice full pitch at low speed, and that could neither generate a cemf or create the low-speed revolving field. If we used a coil span equal to full pitch at low speed (half of full pitch at high speed), the motor behaves as though it was either running at less than rated voltage at low speed or else above rated voltage at high speed. The designer has to choose the coil span carefully to get balanced performance at both speeds.

For consequent-pole motors, NEMA and EEMAC standards require that the schematic be drawn in the high-speed configuration and the leads be numbered so that 1, 2, and 3 will be the low-speed line terminals, and 4, 5, and 6 will be the high-speed line terminals. It is further required that the direction of rotation with the applied voltage phase sequence of 1, 2, 3 shall be the same as that obtained with the phase sequence 4, 5, 6. For this reason if 1, 2, and 3 appear in clockwise order on the schematic, 4, 5, and 6 must appear in counterclockwise order.

FIGURE 4-38 (left)   Two-speed variable-torque motor

MATHEMATICAL INTRODUCTION TO POLYPHASE INDUCTION MOTORS

Coil Pitch and Coil Distribution

It is desirable to have the rotating magnetic field sinusoidally distributed along the air gap at any instant of time. If the flux is not sinusoidally distributed, it can be shown mathematically that the field has harmonic components that rotate at subsynchronous speeds, and some of which rotate backward. These harmonic components of the field flux induce currents in the squirrel cage, causing both a reduction of torque and unnecessary heating, and the forward rotating components create objectionable dips in the torque-speed curve, so the harmonics should be eliminated as nearly as possible.

The main reason for using coil spans of less than 180° is that doing so reduces the harmonic components in the rotating field. Reduced pitch also reduces the fundamental component (which is the only component we want); but, in general, the reduction of harmonics is greater than the decrease of the fundamental components.

The relative effectiveness of a coil is known as the pitch factor (K_p) and is equal to the sine of half the coil span, as given in Equation 4-9.

$$K_p = \sin \frac{Y_s}{2} \qquad (4\text{-}9)$$

where Y_s = coil span in electrical degrees or radians

If the coil span is given as a number of slots, use Equation 2-4 to change to electrical degrees.

The use of more than one coil per pole-phase group also reduces harmonic components in the rotating field, and, again, the fundamental component is reduced, but not as much as the harmonics.

The relative effectiveness of the pole-phase group (compared to having only one coil) is known as the distribution factor (K_d). Some representative values of K_d are given in Table 4-1, or its value can be calculated from Equation 4-10.

$$K_d = \frac{n\alpha/2}{n \sin (\alpha/2)} \qquad (4\text{-}10)$$

where n = number of coils per pole-phase group

α = angular distance between adjacent slots (electrical degrees)

For large values of n, K_d approaches the minimum value of 0.95493 for 60° phase belts and 0.82699 for the 120° belts found in consequent-pole windings.

Pitch and distribution factors of less than 1.0 have the same effect as reducing the number of turns in the coils.

Production of the Rotating Magnetic Field

The following discussion will show mathematically how a rotating magnetic field is produced in a three-phase motor. Consider a three-phase stator like that shown in Figure 4-39. We will use the letter θ to denote an angular position around the stator bore measured in (mechanical) radians counterclockwise from the zero line. If the motor has (P) poles, then $P\theta/2$ is the position in electrical radians.

Let us identify the three phase windings as A, B, and C, and assume that the mmf of

TABLE 4-1   DISTRIBUTION FACTORS FOR 60° PHASE BELTS

Coils per Pole-Phase Group	K_d
1	1.0
2	0.96593
3	0.9598
4	0.95766
5	0.95668
6	0.95614

FIGURE 4-39   Stator: denotes angular position

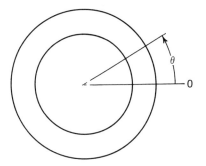

each phase is sinusoidally distributed around the stator bore. We will position phase B a distance of $2\pi/3$ electrical radians counterclockwise from A, and C will then be $(4\pi/3)$ electrical radians ccw from A.

If A is positioned so that its mmf is proportional to $\sin (P\theta/2)$, the mmf of phase B will be proportional to $\sin (P\theta/2 - 2\pi/3)$, and phase C will be proportional to $\sin (P\theta/2 - 4\pi/3)$.

Let us assume that the phase sequence of the currents is A, B, C so that if $i_A = I_{max} \sin \omega t$ (where $\omega = 2\pi f$), then $i_B = I_{max} \sin (\omega t - 2\pi/3)$ and $i_c = I_{max} \sin (\omega t - 4\pi/3)$. The mmf of phase A then is given by Equation 4–11.

$$\text{mmf (phase } A) = K \sin \omega t \sin \frac{P\theta}{2} \quad (4\text{--}11)$$

where K = a constant depending on the number of turns of wire, the value of I_{max}, etc.

The mmf of the whole winding is given by Equation 4–12.

$$\text{mmf total} = K\left[\sin \omega t \sin \frac{P\theta}{2}\right.$$

$$+ \sin \left(\omega t - \frac{2\pi}{3}\right)$$

$$\times \sin \left(\frac{P\theta}{2} - \frac{2\pi}{3}\right)$$

$$+ \sin \left(\omega t - \frac{4\pi}{3}\right)$$

$$\left.\times \sin \left(\frac{P\theta}{2} - \frac{4\pi}{3}\right)\right] \quad (4\text{--}12)$$

The trigonometric identities given in Equations 4–13 and 4–14 are found in most mathematical handbooks, and from them we can derive Equation 4–15.

$$\cos (A - B) = \cos A \cos B + \sin A \sin B$$
$$(4\text{--}13)$$

$$\cos (A + B) = \cos A \cos B - \sin A \sin B$$
$$(4\text{--}14)$$

$$\sin A \sin B = \frac{\cos (A - B) - \cos (A + B)}{2}$$
$$(4\text{--}15)$$

If we apply the identity in Equation 4–15 to the three terms in Equation 4–12, we can write Equation 4–16 and then rearrange the terms as shown in Equation 4–17.

mmf total =

$$\frac{K}{2}\left[\cos \left(\omega t - \frac{P\theta}{2}\right) - \cos \left(\omega t + \frac{P\theta}{2}\right)\right.$$

$$+ \cos \left(\omega t - \frac{2\pi}{3} - \frac{P\theta}{2} + \frac{2\pi}{3}\right)$$

$$- \cos \left(\omega t - \frac{2\pi}{3} + \frac{P\theta}{2} - \frac{2\pi}{3}\right)$$

$$+ \cos \left(\omega t - \frac{4\pi}{3} - \frac{P\theta}{2} + \frac{4\pi}{3}\right)$$

$$\left.- \cos \left(\omega t - \frac{4\pi}{3} + \frac{P\theta}{2} - \frac{4\pi}{3}\right)\right]$$
$$(4\text{--}16)$$

mmf total =

$$\frac{K}{2}\left[\cos \left(\omega t - \frac{P\theta}{2}\right)\right.$$

$$+ \cos \left(\omega t - \frac{2\pi}{3} - \frac{P\theta}{2} + \frac{2\pi}{3}\right)$$

$$+ \cos \left(\omega t - \frac{4\pi}{3} - \frac{P\theta}{2} + \frac{4\pi}{3}\right)$$

$$\left.- \cos \left(\omega t + \frac{P\theta}{2}\right)\right.$$

$$- \cos \left(\omega t - \frac{2\pi}{3} + \frac{P\theta}{2} - \frac{2\pi}{3} \right)$$

$$\left. - \cos \left(\omega t - \frac{4\pi}{3} + \frac{P\theta}{2} - \frac{4\pi}{3} \right) \right] \tag{4-17}$$

For any values of θ and t, the sum of the three right-hand terms in Equation 4–17 is zero (cosines of three angles with 120° between them), so we can delete them and obtain Equation 4–18.

$$\text{mmf total} = \frac{K}{2} \left[\cos \left(\omega t - \frac{P\theta}{2} \right) \right.$$

$$\left. + \cos \left(\omega t - \frac{P\theta}{2} \right) + \cos \left(\omega t - \frac{P\theta}{2} \right) \right] \tag{4-18}$$

Equation 4–18 simplifies to Equation 4–19. If we move along the air gap so that $P\theta/2 = \omega t$, the cosine term remains unity. The mmf we will observe is just one and a half times the mmf of one phase rotating counterclockwise one electrical space radian for each electrical radian of time.

$$\text{mmf total} = \frac{3k}{2} \cos \left(\omega t - \frac{P\theta}{2} \right) \tag{4-19}$$

Voltage, Flux, and Effective Turns Ratio

If the flux is sinusoidally distributed and B_{max} denotes the highest flux density, the average density in the air gap is just $2B_{max}/\pi$. The total air gap flux (ϕ) is just the average flux density times the peripheral area of the stator bore as given by Equation 4–20.

$$\phi = \left(2 \frac{B_{max}}{\pi} \right) (\pi \, dl) \tag{4-20}$$

$$\therefore \phi = 2dlB_{max}$$

where d = diameter of the armature in meters

l = length of the core in meters

B_{max} = maximum flux density in teslas

The average voltage induced in one stator conductor is simply equal to ϕ times synchronous speed in revolutions per second, and its effective value is just 1.11 times average, as shown by Equation 4–21.

$|E|$ (for 1 conductor)

$$= 1.11\phi \cdot (\text{rev/s}) \tag{4-21}$$

The cemf per phase in the stator winding is therefore given by Equation 4–22.

$$|E_s| = \frac{1.11 K_p K_d N_s \phi \cdot (\text{rev/s})}{3b_s} \tag{4-22}$$

where N_s = total number of stator conductors

N_s = 2 × (turns per coil) (total number of coils)

b_s = number of parallel paths per phase

rev/s = synchronous speed

If E_{BA} is the blocked rotor voltage per bar for a squirrel cage or per phase for a wound-rotor motor, Equations 4–21 and 4–22, respectively, can be used to calculate E_{BA} for these machines, but we must be sure to use rotor values for K_p, K_d, and N where they appear.

Let us define the effective turns ratio for a motor as $|E_s|/|E_{BA}|$ and denote it with the letter a. For a squirrel-cage machine,

$$a = \frac{K_{ps} K_{ds} N_A}{3b} \tag{4-23}$$

For a wound rotor,

$$a = \frac{K_{PA}K_{dA}N_A b_s}{K_{ps}K_{ds}N_s b_A} \qquad (4\text{-}24)$$

where the subscript A denotes the rotor quantities and the subscript s indicates stator quantities.

The effective turns ratio in a wound-rotor motor can be determined by measuring the applied voltage on the stator winding (be sure to reduce it to a per phase value) and dividing it by the rotor voltage per phase under blocked-rotor and open-circuit conditions.

Actual and Reflected Rotor Currents

Based upon the principle of equal ampere conductors, we can write Equations 4–25 and 4–26. For a squirrel cage,

$$I_A N_A = \frac{I_R K_{ps}K_{ds}N_s}{b_s} \qquad (4\text{-}25)$$

so that

$$I_R = \frac{I_A N_A}{3a} \qquad (4\text{-}26)$$

For a wound rotor, we can write Equation 4–27, which reduces to Equation 4–28.

$$I_A \frac{K_{PA}K_{dA}N_A}{b_A} = I_R \frac{K_{ps}K_{ds}N_s}{b_s} \quad (4\text{-}27)$$

$$I_R = \frac{I_A}{a} \qquad (4\text{-}28)$$

Synchronous Speed and No-Load Current

Since synchronous speed is inversely proportional to the number of poles, it is ap-

parent from Equation 4–27 that, if we increase the number of poles in a motor and the other factors remain constant, N must increase in the same proportion as p, and so the number of turns per pole remains unchanged. Furthermore, looking at the magnetic circuit, it is apparent that to obtain the same flux density the ampere turns per pole must remain the same, and so the no-load current must remain the same. However, the low-speed motor will be wound with smaller wire; therefore, its permissible full-load current is smaller. For this reason the no-load current of the lower-speed motor will be a greater proportion of its full-load rating. Since the no-load current is nearly always at a rather low lagging power factor, the low-speed motor usually also has a lower power factor and lower efficiency.

Torque of an Induction Motor

The instantaneous force on one rotor conductor is given by Equation 2–6. If we assume sinusoidal flux distribution, constant slip, and unity rotor power factor, the current will also vary sinusoidally with respect to space, and the average force on one conductor is then given by Equation 4–29.

$$F_{avg} = \tfrac{1}{2}B_{max}lI_{A\,max} \qquad (4\text{-}29)$$

The torque developed by one armature conductor is just the product of force times radius, and the total torque on the armature is just the sum of the torques produced by the individual conductors. Rotor power factor must also be taken into account, and so for a squirrel cage we have Equation 4–30, which simplifies to Equation 4–31.

$$T = \tfrac{1}{2}B_{max}lI_{A\,max}N_A\frac{d}{2}\cos\theta_A \quad (4\text{-}30)$$

$$T = \tfrac{1}{4} B_{max} l I_{A\,max} N_A d \cos \theta_A \quad (4\text{–}31)$$

For a wound rotor, we need only to insert pitch, distribution factors, and the number of parallel paths to get Equation 4–32.

$$T = \frac{1 K_{PA}}{4 b_A} K_{dA} B_{max} l I_{A\,max} N_A d \cos \theta_A \quad (4\text{–}32)$$

Equations 4–31 and 4–32 are the fully developed versions of Equation 4–7, and the striking similarity between these equations and Equation 1–10 should also be noted.

Equivalent Circuit

The equivalent circuit for an induction motor can be derived from basic physical considerations, and much of the work has already been done in previous sections of this chapter.

Since $I_A = E_A/(R_A + jX_A)$, $E_A = s|E_s|/a$, and $X_A = sX_{BA}$, we can write Equation 4–33.

$$I_A = \frac{s|E_s|/a}{R_A + jsX_{BA}} \quad (4\text{–}33)$$

Since $I_R = I_A/a$, we can obtain Equation 4–34 for a wound-rotor motor.

$$I_R = \frac{s|E_s|}{a^2(R_A + jsX_{BA})} \quad (4\text{–}34)$$

Equation 4–34 can be further modified to get Equation 4–35.

$$I_R = \frac{|E_s|}{a^2(R_A/s) + ja^2 X_{BA}} \quad (4\text{–}35)$$

If we let $a^2 R_A = R_R$ and $a^2 X_{BA} = X_R$, we can write Equation 4–36.

$$I_R = \frac{|E_s|}{(R_R/s) + jX_R} \quad (4\text{–}36)$$

For a squirrel-cage motor, $I_R = I_A N_A/3a$, so from Equation 4–33 we can derive Equation 4–37.

$$I_R = \frac{s|E_s|/a}{R_A + jX_{BA}} \cdot \frac{N_A}{3a} \quad (4\text{–}37)$$

$$\therefore I_R = \frac{s|E_s|N_A}{3a^2(R_A + jsX_{BA})}$$

We can further modify Equation 4–37 to get Equation 4–38.

$$I_R = \frac{E_s}{(a^2 R_A/N_A s) + (ja^2 X_{BA}/N_A)} \quad (4\text{–}38)$$

If we let $R_R = 3a^2 R_A/N_A$ and $X_R = 3a^3 X_{BA}/N_A$, and substitute these values into Equation 4–34, we again obtain Equation 4–36.

Equation 4–36 tells us that the reflected rotor currents for an induction motor can be exactly simulated by connecting the impedance $(R_R/s + jX_R)$ in parallel with the stator winding. R_R is known as the reflected rotor resistance, and X_R is known as the reflected rotor reactance. Although their numerical values are obtained in slightly different ways, these quantities apply in the same fashion to both squirrel-cage and wound-rotor machines. It is interesting to note that, while the actual rotor reactance (X_A) and rotor voltage (E_A) are proportional to slip and R_A is constant, the equivalent circuit has constant voltage $(|E_s|)$, constant reactance (X_R), and resistance (R_R/s) that varies inversely with the slip.

The exciting current (I_o) that flows through the stator windings has two components, a magnetizing component, which is 90° behind the voltage and an in-phase component that makes up the eddy currents and hysteresis losses in the stator. These can be incorporated into our equivalent circuit by means of a conductance (G_o) and a susceptance (B_o) of appropriate values. The resistance and leakage reactance of the stator winding (R_s and X_s) quite naturally belong in series with all the other portions of the circuit; so we end up with the equivalent circuit shown in Figure 4-40. It is understood that this circuit represents just one phase of the stator winding. If the stator is delta connected, the line voltage applied to the motor is also the voltage applied to this circuit (E_s), and the line current is just the current in the equivalent circuit ($|I_s|$) multiplied by $\sqrt{3}$. If the stator is wye connected, the voltage applied to the equivalent circuit is just line voltage divided by $\sqrt{3}$, and the current in the equivalent circuit is the actual line current. However, no matter how the stator is connected, power values calculated from the equivalent circuit must be multiplied by 3 to get the total for the whole motor. Stator copper loss is three times the power in R_s, stator iron loss is three times the power in G_o, and rotor power input (RPI) is three times the power in R_R/s.

For a squirrel cage, the rotor copper loss equals $|I_A|^2 R_A N_A$, and from Equation 4-26 we can see that I_A equals $3aI_R/N_A$. We can therefore derive Equation 4-39 as follows.

Rotor copper loss
$$= \frac{9a^2|I_R|^2}{N_A^2} \cdot R_A N_A$$

$$= \frac{3|I_R|^2(3a^2 R_A)}{N_A}$$

$\therefore$ Rotor copper loss
$$= 3|I_R|^2 R_R \qquad (4\text{-}39)$$

Equation 4-39 is also valid for a wound-rotor machine.

We can find the rotor power developed either by subtracting rotor copper loss from rotor power input or by using Equation 4-40.

$$\text{RPD} = (1 - s)\text{RPI} \qquad (4\text{-}40)$$

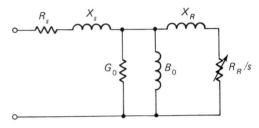

FIGURE 4-40 Equivalent circuit for an induction motor

If RPD and the rotational speed of the rotor are known, we can calculate the torque using Equation 4-41.

$$P = 2\pi \, T(\text{rev/s}) \qquad (4\text{-}41)$$

However, since the rotor speed and rotor power developed are both proportional to $(1 - s)$, we can write Equation 4-42. Since the equivalent circuit does not take them into account, the mechanical losses must be subtracted to obtain the net torque or power output.

$$T = \frac{\text{RPI}}{2\pi} \cdot (\text{syn. rev/s}) \qquad (4\text{-}42)$$

Breakdown torque always occurs when we have maximum power dissipation in R_R/s. For most motors, we can approximate the slip at which breakdown torque occurs using Equation 4-43. Equation 4-44 gives us the approximate breakdown RPI.

$$s = \frac{R_R}{\sqrt{R_s^2 + (X_s + X_R)^2}} \qquad (4\text{-}43)$$

$$\text{RPI} = \frac{3E_s}{2\sqrt{R_s^2 + (X_s + X_R)^2}} \qquad (4\text{-}44)$$

The Steinmetz equivalent circuit is an excellent design tool because it permits calculation of motor performance using little more than basic ac theory. Unfortunately, it is almost impossible to obtain the required values for this equivalent circuit through laboratory measurements taken on an existing motor.

Laboratory Tests on Induction Motors

There are four basic laboratory tests that can be conveniently done on either squirrel-cage or wound-rotor motors. There are other useful tests that can be performed on wound-rotor machines, but we will not discuss them here.

There is no need to determine the actual winding connections. We can assume a wye connection without introducing any error, and our mathematics in this and the next section is based on that assumption.

The four tests are the following:

Stator Resistance Test

To obtain the stator winding resistance, we must do some kind of stator resistance test. An ohmmeter of suitable range may be used, but if such is not available, use a dc power supply and a voltmeter–ammeter technique. The dc resistance per phase is just half the terminal-to-terminal resistance that we normally obtain in this test. To get the effective ac resistance per phase, multiply the dc resistance per phase by 1.5.

No-Load Test

The no-load test consists of running the motor at rated voltage and no load and recording the voltage, current, and total power input. The no-load speed is usually within 0.05 rev/s of synchronous speed and we usually ignore this difference. For some purposes it may be desirable to obtain a measurement of slip (if it can be directly obtained) or a very accurate measurement of rotor speed.

Measurements Under Load

Measurements under load generally include voltage, current, total power input, and slip (or speed) with a mechanical load on the motor. Sometimes we take several sets of readings at various loads; for some purposes we may wish to measure the motor torque while it is loaded.

Blocked-Rotor Test

For the blocked-rotor test we prevent the armature from turning, apply a balanced three-phase voltage, and measure the voltage, current, and total power input. For some purposes it may be desirable to measure the torque developed during the test. We almost always do this test at less than half the voltage rating of the motor. To find the measurements that would be obtained if the blocked-rotor test were done at rated voltage (we will call these the corrected measurements), remember that current is proportional to voltage, while power and torque are both proportional to voltage squared. Example 4-1 shows what to do.

EXAMPLE 4-1

Given the following data from the blocked-rotor test on a 230-V motor, determine the corrected values.

Line voltage = 90 V

Line current = 35 A

Total power input = 3.7 kW

Torque = 12 newton-meters (N · m)

Solution

The corrected values are

$$|V| = 230 \text{ V} \quad \text{(by definition)} \quad \text{(Answer)}$$

$$|I| = 35 \times \frac{230}{90} = 89.44 \text{ A} \quad \text{(Answer)}$$

Total power $= 3.7 \times \left(\dfrac{230}{90}\right)^2$

$\qquad = 24.16$ kW $\qquad$ (Answer)

Torque $= 12 \times \left(\dfrac{230}{90}\right)^2$

$\qquad = 78.37$ N · m $\qquad$ (Answer)

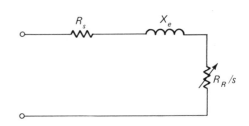

FIGURE 4-41   Series equivalent circuit for an induction motor

Motor Performance from Laboratory Tests

If we wish to calculate motor performance using data obtained from the preceding laboratory tests, we will be forced to make some approximations. To obtain good accuracy, the mathematical procedures can get rather lengthy and difficult to remember. The following procedures have been included here because they are accurate enough for teaching purposes, and they can be easily recalled by looking at the associated diagram. We will always assume wye connections.

Approximate Equivalent Circuits

These circuits are usable for wound-rotor motors and single squirrel-cage types if the current-displacement effect is negligible. Before starting any mathematics, be sure that all data are given on a per phase basis. Phase voltage equals (line voltage)/$\sqrt{3}$; power and torque per phase are just one third of the total values. The phase and line currents are identical. There are two convenient equivalent circuits.

SERIES EQUIVALENT CIRCUIT   To derive the series equivalent circuit, we assume that the no-load current is negligible. If so, then I_o is zero, B_o and G_o disappear from the

equivalent circuit in Figure 4-40, and we end up with the circuit in Figure 4-41. The reactance (X_e) is basically the sum of X_s and X_R, and to get numerical values for this circuit, all we need is the blocked-rotor test data including torque. Example 4-2 shows the procedure. The series equivalent circuit is not of much value for anything but torque calculations.

EXAMPLE 4-2

Draw a series equivalent circuit for one phase of the motor in Example 4-1. Find the ohmic values for R_s, X_e, and R_R, and assuming a synchronous speed of 30 rev/s, find the motor torque at rated voltage and 5% slip.

Solution

On a per phase basis the blocked rotor test data are as follows:

$$|V| = \frac{90}{\sqrt{3}} = 52 \text{ V}$$

$$|I| = 35 \text{ A}$$

$$P = \frac{3.7}{3} = 1.233 \text{ kW}$$

$$T = \frac{12}{3} = 4 \text{ N} \cdot \text{m}$$

From Equation 4-42,

$$4 = \frac{\text{RPI}}{2\pi(30)}$$

$$\therefore \text{RPI} = 2\pi \times 30 \times 4 = 754 \text{ W}$$

Since $P = I^2 R$, and $s = 1$ during the test,

$$R_R = \frac{754}{35^2}$$

$$= 0.6155 \ \Omega \quad \text{(Answer)}$$

$$\text{Total } R = \frac{1233}{35^2}$$

$$= 1.0068 \ \Omega$$

$$R_s = 1.0068 - 0.6155$$

$$= 0.3913 \quad \text{(Answer)}$$

$$|Z| = \frac{52}{35}$$

$$= 1.486 \ \Omega$$

$$\therefore X_e = \sqrt{1.486^2 - 1.0068^2}$$

$$= 1.093 \ \Omega \quad \text{(Answer)}$$

$$\frac{R_R}{s} = \left(\frac{0.6155}{s}\right) \Omega$$

If $s = 0.05$,

$$\frac{R_R}{s} = \frac{0.6155}{0.05} = 12.31 \ \Omega$$

$$\therefore R_{\text{total}} = 12.31 + 0.3913 = 12.7013 \ \Omega$$

$$|Z|_{\text{total}} = \sqrt{12.7013^2 + 1.0068^2} = 12.74 \ \Omega$$

At rated voltage,

$$|V| = \frac{230}{\sqrt{3}} = 133 \ V$$

$$|I| = \frac{133}{12.74} = 10.42 \ A$$

$$\text{RPI} = 10.42^2 \times 12.31$$

$$= 1337 \ W \text{ per phase}$$

$$\text{Total RPI} = 1337 \times 3 = 4012 \ W$$

$$\text{Total } T = \frac{4012}{2\pi(30)}$$

$$= 21.3 \ N \cdot m \quad \text{(Answer)}$$

L-SHAPED EQUIVALENT CIRCUIT This circuit is derived by moving the magnetizing branch (G_o and B_o) to the line side of R_s

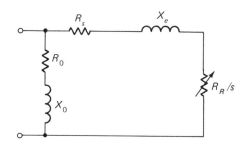

FIGURE 4-42   L-shaped equivalent circuit for an induction motor

in Figure 4-39. If we are not contemplating a change of frequency, G_o and B_o in parallel can be replaced with R_o and X_o in series, as shown in Figure 4-42. Again $X_e = X_s + X_R$. Almost anything (such as I_s, efficiency, etc.) can be calculated from this circuit with reasonable accuracy except for stator copper loss, which is not well represented.

To derive an L-shaped equivalent circuit, we need a no-load test (do not worry about slip), and we would prefer a blocked-rotor test that includes a torque measurement. If torque cannot be measured, a stator resistance test must also be performed.

Before starting any other mathematics, reduce the available data to per phase values, and the blocked-rotor data must be corrected values on a per phase basis. In addition, change the currents from absolute values (meter readings) to complex (j operator) notation.

Under no-load conditions, s becomes very, very small, and so the rotor branch of the equivalent circuit (where I_R is found) becomes essentially an open circuit. The values of R_o and X_o can therefore be found using the same procedure as in Example 4-2.

Using complex notation, subtract the no-load current from the blocked-rotor current (which gives I_R); also subtract the no-load power from the blocked-rotor power (which gives the total power in R_R and R_s. Get the absolute value of I_R, and we now have enough information to find X_e and the sum of R_R and R_S. If a torque measurement is available, find R_R and then get R_S by subtraction. If R_S is known find R_R by subtraction.

It is interesting to note that windage and friction as well as iron losses are included in R_o of this circuit, and that the sum of these rotational losses is assumed to be constant.

Power-Flow Diagrams

What we basically require is a no-load test and one or more sets of measurements of voltage current, power input, and slip (or speed) under load. In addition, we must do either a stator resistance test or a blocked-rotor test without a torque measurement so that we can proceed in one of two ways:

SEGREGATED COPPER LOSS CALCULATIONS If the stator resistance is known, we can find the stator copper loss at any value of current ($3I_S^2 R_S$). If we subtract the no-load copper loss from the no-load input, we then have the rotational losses. It is convenient to assume that these will be constant and charge them to the rotor. We are now also in a position to calculate copper losses at any value of stator current. If at any value of I_S we subtract the stator copper loss, we will have the rotor power input. If the slip or speed is known, we can easily find rotor power developed using Equation 4–40. Subtracting the rotational losses then gives us the power output. Example 4-3 shows what to do.

In this procedure we are treating the iron losses as if they were mechanical so that the power-flow diagram of Figure 4-18 becomes that shown in Figure 4-43. The procedure

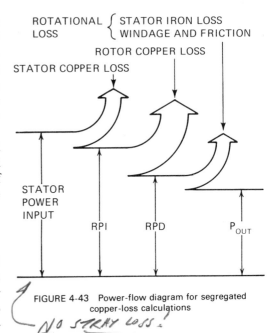

FIGURE 4-43   Power-flow diagram for segregated copper-loss calculations

NO STRAY LOSS!

is equally valid for double squirrel-cage machines or one where current displacement is significant.

EXAMPLE 4-3

At no load, a certain motor draws 5 A at 460 V, and the power input is 900 W. The effective ac resistance of the stator is 0.95 Ω per phase. Find the power output if the motor is running at 94% of synchronous speed, the current is 19 A, and the power input is 6.9 kW.

Solution

No-load copper loss $= 3 \times 5^2 \times 0.95$
$$= 71.25 \text{ W}$$

$\therefore$ Rotational losses $= 900 - 71.25$
$$= 828.75 \text{ W}$$

At 19 A,

Stator copper loss $= 3 \times 19^2 \times 0.95$
$$= 1029 \text{ W}$$

$\therefore$ RPI $= 6.9 - 1.029$
$$= 5.871 \text{ kW}$$

RPD = 5.871 × 0.94

= 5.519 kW

Less rotational losses

≅ 0.829

Power output

= 4.690 kW   (Answer)

NONSEGREGATED COPPER LOSS CALCULA-
TION   Under blocked-rotor conditions, the
mechanical losses are zero, and the iron
losses are somewhat depressed compared to
no-load conditions. If we assume their sum
to be zero at zero speed, all the blocked-
rotor power input must be copper losses.
Based upon the series equivalent circuit,
we can determine the sum of R_S and R_R, and
this sum can be used to find total copper
losses at any value of current; so we can
easily find the mechanical losses from the
no-load data. Subtracting the copper losses
from the input gives us RPD, and if we then
subtract the mechanical losses, we will have
the power output. Example 4-4 illustrates
the procedure.

In this method we are ignoring the existence of the magnetizing current (I_o). We
therefore can lump stator and rotor copper
loss together so that the power-flow diagram
in Figure 4-43 becomes that in Figure 4-44.
This procedure is not very accurate for motors
with double squirrel cages or substantial
current-displacement effect.

EXAMPLE 4-4

If the motor from Example 4-2 draws 4 A and
300 W at 230 V and no load, find its power
output at 230 V, 10.4 A, 3.1 kW, and 5% slip.

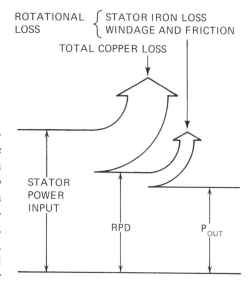

FIGURE 4-44   Power-flow diagram for
nonsegregated copper-loss calculations

Solution

$$\text{Total resistance} = \frac{1233}{35^2} = 1.0068$$

At no load,

Total copper loss = 3 × 4² × 1.0068

= 48.33 W

∴ Rotational losses = 300 − 48.33

= 251.67 W

At 10.4 A,

Total copper loss = 3 × 10.4² × 1.0068

= 325.7 W

∴ RPD = 3100 − 325.7

= 2774.3 W

∴ Power output = 2774.3 − 251.67

= 2572.63 W   (Answer)

SUMMARY

The three-phase squirrel-cage motor is the
type most commonly used by industry today.

The operation of an induction motor is fairly
easy to visualize and can be understood in

terms of elementary physical principles. To get a reasonable mathematical description, we used about 40 equations in this chapter, but the Steinmetz equivalent circuit is a relatively simple way to represent the motor, and the mathematics required to solve that circuit is quite elementary.

The most complex part of a squirrel-cage motor is the stator windings. However, they follow one of a rather few distinct patterns, and once the pattern is recognized it becomes relatively simple. The multispeed motor windings are probably the most difficult to understand.

Because they are a high-volume production item, induction motor designs have been refined to a very high degree over the years. The material in this chapter is only the barest introduction to induction motor design. Readers who wish to pursue that subject in greater detail are encouraged to seek out other sources of information.

REVIEW QUESTIONS

4-1. Explain what is meant by the following terms:
  (a) stator
  (b) skewed slots
  (c) semiclosed slots
  (d) random-wound coils
  (e) rotor
  (f) squirrel cage
  (g) synchronous speed
  (h) slip
  (i) rotor power factor
  (j) current displacement
  (k) double squirrel cage
  (l) induction generator
  (m) slip-ring motor
  (n) secondary circuit of a motor
  (o) coil span
  (p) 1 to 7 connection
  (q) electrical space degree
  (r) pole-phase group
  (s) dual-voltage motor
  (t) triple-power-rated motor
  (u) wye start–delta run motor
  (v) consequent-pole windings
  (w) pole-changing winding
  (x) multispeed motor
  (y) constant-power motor

4-2. Draw a diagram showing the magnetic paths in a six-pole motor.

4-3. Explain how the rotating magnetic field is produced in a three-phase motor.

4-4. What is the easiest way to reverse the rotation of a three-phase motor?

4-5. Explain how a squirrel cage develops torque.

4-6. Why do induction motors have a very definite no-load speed?

4-7. Explain why the following rotor quantities are proportional to slip:
  (a) voltage
  (b) frequency
  (c) reactance

4-8. Explain why torque is reduced when the rotor current lags the rotor voltage.

4-9. Increasing the rotor resistance does not reduce breakdown torque, but merely reduces the speed at which it occurs. Explain why.

4-10. What is meant by "reflected rotor current" and why does it occur?

4-11. Explain why current displacement occurs, and show with a diagram how it affects the torque–speed curve.

4-12. What is a double squirrel-cage rotor and why is it used?

4-13. What is meant by regenerative braking, and how can it occur in squirrel-cage motors?

4-14. If we interchange two of the three wires connected to the brushes of a wound-rotor motor, how will this affect rotation and why?

4-15. What are the three main advantages of wound-rotor motors?

4-16. What are the four losses of an induction motor?

4-17. Why are rotor iron losses negligible under normal running conditions?

4-18. What is the most obvious difference between a conventional winding and a basket winding?

4-19. What is meant by the top and bottom sides of a coil?

4-20. Which pole-phase groups belong to phase *A* of an eight-pole motor?

4-21. Draw a diagram (similar to Figure 4-20) showing the following motor winding:

$$Number\ of\ slots\ =\ 24$$
$$Number\ of\ coils\ =\ 24$$
$$Coil\ span\ =\ 5$$

Number of pole-phase groups = 12

4-22. A six-pole motor is to be connected series wye using a 1 to 4 connection. Draw a diagram like that in Figure 4-22 to show how this is done.

4-23. Using your diagram for question 4-22, explain how the rotating field is produced.

4-24. A certain manufacturer builds a 96-slot stator and installs conventional windings for 2, 4, 6, 8, 10, or 12 poles as required. For each of these windings, determine the number of coils per pole-phase group. If nonuniform groups will result, specify how many coils there are in each size of group, how many groups of each size will be required, how they are distributed around the stator, and the possible numbers of parallel paths for which the winding could be connected.

4-25. What is the main advantage of a dual-voltage motor?

4-26. A certain 12-pole motor with uniform coil groups is series wye connected and rated at 2300 V. If we take the motor apart so that the intergroup connections are accessible, for what other voltages could this motor be connected? (*Hint:* As long as the voltage per turn remains the same, motor performances will be unchanged.)

4-27. Draw schematic diagrams for dual-voltage wye- and delta-connected motors. Number the leads according to EEMAC standards and provide connecting instructions for each.

4-28. A motor rated at 20/15/10 kW output has 2.0 V per turn when connected for 20 kW. What should the voltage per turn be on the other two connections? (*Hint:* Torque is proportional to the square of the voltage per turn.)

4-29. Draw a schematic diagram of a triple-power-rated motor. Identify the leads according to NEMA standards and provide connecting instructions for each power rating.

4-30. Draw the following schematic diagrams and identify the leads according to EEMAC standards.
    (a) Motor for wye-delta starting.
    (b) Motors for part winding starting (three diagrams).

4-31. Using the hint from question 4-28, explain why wye–delta starting reduces torque to one third of the normal values.

4-32. Assuming constant impedance per phase, explain why wye–delta starting reduces starting current to one third of normal.

4-33. What is meant by part winding starting and why is it used?

4-34. Draw a diagram like that in Figure 4-33 for a six-consequent-pole motor using a 1 to 7 connection and starting on groups 1, 3, and 5.

4-35. Using your diagram for question 4-34, explain how a rotating magnetic field is produced by a consequent-pole motor winding.

4-36. How do pole-changing windings accomplish the change?

4-37. Distinguish between constant-power, constant-torque and variable-torque motors.

4-38. If a motor has two stator windings, it is preferable that both of them be series-wye connected. Why?

4-39. Draw wiring diagrams for all three types of two-speed consequent-pole motors. Make them two- to four-pole machines; number external leads in accordance with NEMA standards.

4-40. Draw schematic diagrams for all three types of two-speed consequent-pole motors. Identify the leads in accordance with NEMA standards and give the connecting instructions for each.

MATHEMATICS PROBLEMS

4-1. Make a table that shows synchronous speeds for all motors from 2 to 14 poles inclusive when operated at 25, 50, 60, 400, and 800 Hz.

4-2. Find the full-load slip (in percent) for each of the following motors. The speeds quoted are rotor speeds at full load.
(a) 2 pole, 60 Hz, 59.2 rev/s
(b) 4 pole, 50 Hz, 24.1 rev/s
(c) 8 pole, 400 Hz, 99 rev/s
(d) 12 pole, 60 Hz, 8.5 rev/s

4-3. Find the rotor frequency for each motor in problem 4-2.

4-4. Under locked-rotor conditions, a certain squirrel-cage motor has a locked-rotor voltage of 800 mV, and at rated voltage its torque equals $0.2511 I_A$ cos θ_A. Calculate the rotor current and torque at 1, 2, 5, 10, 20, 50, 100, and 200% slip for each of the following rotors and graph the results (similar to Figures 4-8 and 4-11).
(a) $R_A = 1$ mΩ/bar
$X_{BA} = 10$ mΩ/bar
(b) $R_A = 2$ mΩ/bar
$X_{BA} = 10$ mΩ/bar
(c) $R_A = 1$ mΩ/bar
$X_{BA} = 5$ mΩ/bar
Note: It is easiest to make three tables for this problem, each with a row for each value of slip and a column for each of the quantities s, E_A, R_A, X_A, Z_A, I_A, cos θ_A, and T. Now fill in the tables one column at a time.

4-5. Find the power input, stator copper loss, stator iron loss, RPI, rotor copper loss, mechanical losses, RPD, efficiency, and/or power output as appropriate for each of the following motors.

(a) Stator copper loss = 3.2 kW
Rotor copper loss = 2.9 kW
Power input = 95 kW
Stator iron loss = 1.1 kW
Mechanical loss = 1.3 kW

(b) Stator copper loss = 500 W
Rotor copper loss = 600 W
Power output = 5 kW
Stator iron loss = 200 W
Mechanical loss = 275 W

(c) Stator copper loss = 10 kW
Rotor copper loss = 11.5 kW
Efficiency = 87%
Stator iron loss = 4.5 kW
Mechanical loss = 8 kW

(d) Power input = 29 kW
Stator iron loss = 1.3 kW
Rotor power input = 25 kW
Rotor copper loss = 2.9 kW
Power output = 21 kW

4-6. Find the per unit slip of each motor in question 4-5.

4-7. Referring to the triple-power-rated motor in Figure 4-30, let us assume that the taps are at the one-quarter and three-quarter points in each phase, and with the maximum power connection there will be 2 V per turn on the windings. Assuming equal voltage per turn on each winding, find the voltage per turn with the other two connections, and find the torque with those connections as a percentage of the torque when connected for maximum power.

4-8. Consider the three types of two-speed consequent-pole motors. If they each have 2 V per turn on the high-speed connection, find the voltage per turn for each motor on its low-speed connection.

4-9. Find the pitch factor for each of the following motors.

(a) Coil span of 150°, 7 coils per pole-phase group
(b) 96 slots, 8 poles, coil pitch = 11
(c) 48 slots, 4 poles, coil pitch = 10
(d) 48 slots, two-speed 4- to 8-pole, coil pitch = 9 (two answers)
(e) 36 slots, two-speed 2- to 4-pole, coil pitch of 1 + 11 (two answers)

4-10. Find the distribution factor for each motor in problem 4-9 (two answers are required for each two-speed motor).

4-11. Referring to the mathematics on page 129 show mathematically that reversing the sequence of the phase currents will reverse the rotation of the total mmf.

4-12. A two-phase motor is similar to a three-phase type, but it has only two stator windings that are positioned 90 electrical space degrees from each other. Show that if those windings carry equal currents that are 90° out of phase with each other a rotating mmf with a magnitude equal to the peak mmf of one phase will be produced.

4-13. Find the number of turns of wire per coil for each of the stator windings listed below. All windings are for a 36-slot core that is 0.3 m long and takes a rotor 0.25 m in diameter. The voltages quoted are line-to-line applied voltages, and it is assumed that the cemf equals the line voltage.

(a) 2 pole, 60 Hz, 208 V, series wye connected, coil span = 15, B_{max} = 1.1 teslas (T)
(b) 6 pole, 50 Hz, 600 V, series delta connected, coil span = 5, B_{max} = 1.2 T
(c) 4 pole, 25 Hz, 480 V, 2 parallel wye connected, coil span = 10, B_{max} = 1.25 T

4-14. If the motors in problem 4-13 are squirrel-cage types, find the blocked-rotor voltage per bar (E_{BA}) and the effective turns ratio.

4-15. For each of the motors in problem 4-13, we will use a wye-connected wound rotor that has a pitch factor of 0.9, a distribution factor of 0.96, and a total of 840 armature conductors. Find the blocked-rotor voltage per phase (E_{BA}), the blocked-rotor voltage between slip rings, and the effective turns ratio of the motor.

4-16. If the rotors in problem 4-14 each have 31 bars and the rms current in each bar is 300 A, find the reflected rotor current per phase and also per line in the stator.

4-17. If the rotor in problem 4-15 has a current flow of 15 A at each slip ring in each case, calculate the reflected rotor current per phase in each case.

4-18. If the rotor in problem 4-16 has 1.3-mΩ resistance and 17.24-μH inductance per bar, find the reflected rotor resistances and rotor reactance in each case.

4-19. If the rotor in problem 4-17 has 2-Ω resistance and 31.831 mH of inductance per phase, find R_R and X_R for each motor.

4-20. Find the blocked-rotor torque of each rotor in problems 4-18 and 4-19.

4-21. A 208-V, four-pole, 60-Hz, wye-connected motor has the following equivalent circuit values. Find the torque, power output line current, power factor, and efficiency at each of 0, 5, 10, 15, 25, 40, 70, and 100% slip.

$$R_s = 0.5 \ \Omega$$

$$X_s = 1.0 \ \Omega$$

$$B_o = 0.03782 \text{ siemens (S)}$$

$$G_o = 0.009066 \text{ S}$$

$$X_R = 2.0 \ \Omega$$

$$R_R = 0.8 \ \Omega$$

4-22. If a 208-V, four-pole, 60-Hz motor gives the following test results, derive an L-shaped equivalent circuit, and find the line current, torque, power output, power factor, and efficiency at each of 0, 5, 10, 15, 25, 40, 70, and 100% slip.

No-load test at 208 V:

$$I_{\text{line}} = 4.47 \text{ A}$$

$$P_{\text{in}} = 390 \text{ W} \quad \text{(total)}$$

Blocked-rotor test data corrected to 208 V:

$$I_{\text{line}} = 38.6 \text{ A}$$

$$P_{\text{in}} = 5447 \text{ W}$$

$$T = 16.1 \text{ N} \cdot \text{m}$$

4-23. Repeat problem 4-22 assuming that the effective stator resistance is 6 Ω per phase and the blocked-rotor torque is not known.

4-24. Referring to the motor in problem 4-22, if only the blocked-rotor test data are known, derive a series equivalent circuit, and calculate the torque at 5, 10, 15, 25, 40, and 70% slip.

4-25. Referring to the motor in problem 4-23, find its power output if it draws 2.746 kW and 9.346 A when running at 5% slip.

4-26. Assuming that neither the stator resistance nor blocked-rotor torque is known, find the power output of the motor in problem 4-22 if it draws 4.583 kW and 15.15 A when running at 10% slip.

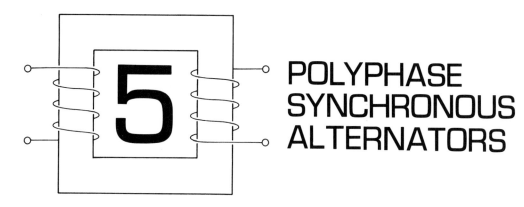

POLYPHASE SYNCHRONOUS ALTERNATORS

An alternator receives energy in mechanical form and converts it to electrical energy. It is quite possible for an individual user to operate an alternator that is adequate for his loads, but in most populated areas it is more economical and convenient to have an electrical utility company that generates and distributes electrical power to all or most of the users in that specific area. Such a utility company can make use of very large alternators, which are inherently more efficient, and for this reason alternators as large as a million kilowatts of output are now in service.

In recent years the growing concern about efficient use of energy has made users reconsider the economics of generating all or part of their own electrical requirements. Hospitals, large educational institutions, industrial plants like oil refineries, and even large office buildings sometimes find it economical to generate their own power, usually in conjunction with their heating equipment.

It seems likely that the number of smaller, user-owned alternators will increase considerably in the forthcoming years.

We will start by examining the construction of alternators. Then we will briefly consider the theory of operation and typical characteristics, with special attention to phasing out, synchronizing, and parallel operation. Then we will briefly describe the special cooling systems used on large alternators and mathematically consider the generated voltage, voltage regulation calculations, the synchronous impedance test, losses and efficiency, and the limits of permissible loading.

Most alternators have rather extensive protection and control systems. Automatic synchronizing equipment, automatic voltage regulators, overcurrent protection, differential protection, and/or split-winding protection schemes are commonly used, but these topics seem too specialized to include in this text.

CONSTRUCTION AND THEORY OF OPERATION

Basic Parts of a Synchronous Alternator

There are two essential components:

Stator Core and Winding Assembly

This unit is very similar to that of an induction motor. An alternator designed for use with a high-speed prime mover (such as a steam or gas turbine) is usually wound for either two or four poles, but if intended to be driven by an engine or a hydraulic turbine, they may have 6 to 60 or more poles. The stator windings may be connected either wye or delta; if a wye connection is used, the wye point may be brought out as an accessible external terminal. Because of their size, large alternators frequently have hydrogen- or liquid-filled cooling systems to carry away the heat.

Rotor Assembly

This is basically a cylindrical iron core mounted on a shaft with an insulated dc winding arranged to produce magnetic poles on its outer periphery. The number of magnetic poles on the rotor is always equal to the number of poles for which the stator has been wound. The rotor is commonly known as the field.

In addition to the dc winding, the rotor may have a squirrel-cage winding known as a damper winding or an amortisseur winding. If the alternator is intended to be driven by a gasoline or diesel engine, it will almost certainly have a damper winding; but if it will be driven by a turbine prime mover (be it hydraulic, steam, or gas), the damper winding may be omitted. The reason for this is that the damper winding squelches any tendency to hunt, and while the torque pulsations of an engine tend to cause hunting, the smooth torque of a turbine does not.

Figure 5-1 shows cross-sectional views of a four- and a six-pole alternator.

No-Load Operation

The flow of dc excitation current through the rotor (field) winding makes the rotor a magnet. When the rotor is turning, its flux

FIGURE 5-1 Four- and six-pole alternator cross sections

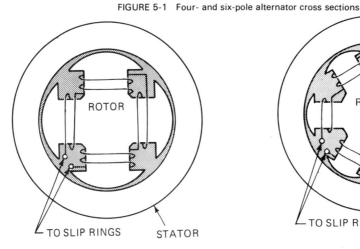
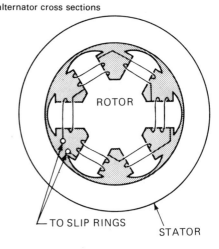

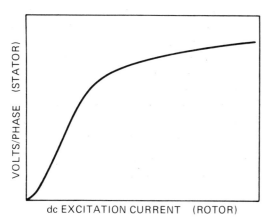

FIGURE 5-2   Typical no-lead saturation curve for an alternator

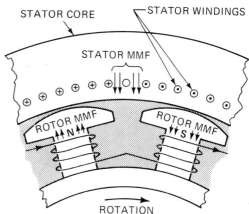

FIGURE 5-3   Armature reaction when the current is in phase with the rotor flux

cuts the stator windings, inducing a voltage there. Since the stator has three identical windings mechanically displaced 120 electrical degrees from each other, the voltages of those windings are equal and 120° out of phase with each other, so a balanced three-phase output is obtained.

At low values of excitation, the voltage of an alternator tends to bear a linear relationship to the excitation current, but at high excitation levels, saturation of the magnetic circuit tends to put a ceiling on the voltage that can be produced.

A graph that shows the no-load ac voltage generated (ordinate) on a base of excitation current (abscissa) is known as the no-load saturation curve of the alternator. See Figure 5-2 and note the similarity to Figure 1-26 for a dc generator.

Armature Reaction

The voltage generated in a synchronous alternator must always be in phase with the air gap flux in the machine, and at no-load that corresponds to the rotor flux. However, with a load on the alternator the flow of

armature current creates an mmf that tends to change the air gap flux, and the effect of this armature reaction needs to be examined in more detail.

Figure 5-3 shows a cross-sectional view of an alternator, and the crosses and dots appearing in the conductor cross sections indicate the instantaneous direction of the voltage induced in each. If the stator current is in phase with the voltage, the same crosses and dots indicate the current directions, too. We can now apply the right-hand rule to determine the direction and position of the stator mmf. It is important to notice that, if the current is in phase with the rotor flux, the stator mmf lags the rotor mmf by 90°.

Figures 5-4 and 5-5 show the position of the stator mmf when the current lags or leads the rotor flux by 90°. In Figure 5-4 the current is lagging by 90° and the stator mmf is directly opposing the rotor mmf; in Figure 5-5 the stator mmf directly adds to the rotor mmf.

From Figure 5-4 we can recognize that armature reaction with a lagging power factor load will lower the output voltage of an

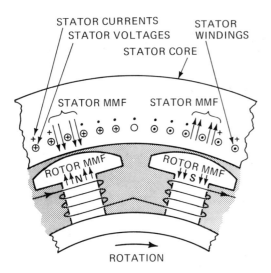

FIGURE 5-4 Armature reaction when the stator
current is 90° behind the rotor flux

alternator more (i.e., cause more voltage regulation) than it will with a unity power factor load.

From Figure 5-5 we can see that with a leading power factor load armature reaction

FIGURE 5-5 Armature reaction when the stator current leads the rotor flux by 90°

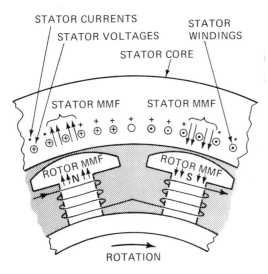

will tend to increase the flux (and therefore the voltage), with the result that the voltage regulation becomes rather small or even takes on a negative value.

Armature reaction in an alternator is a major cause of voltage regulation. However, since it creates a voltage drop that leads the current by 90°, its effect on the voltage is the same as having some inductive reactance in series with the alternator, and that is how we will treat it using phasor diagrams and the appropriate mathematics.

Synchronous Impedance and Voltage Regulation

Voltage regulation in an alternator has three causes:

- Armature reaction
- Leakage reactance of the stator windings
- Stator winding resistance

The voltage drop due to resistance is always in phase with the current, but armature reaction and armature reactance both create a voltage drop that leads the current by 90°.

For voltage regulation calculations with a balanced load, it is easiest to view the alternator as a constant-voltage source with an impedance known as synchronous impedance (Z_s) in series with it in each phase. The resistance portion of it (R_A) is the effective ac resistance of the stator windings, generally about 1.5 times the dc resistance. The reactance portion is called the synchronous reactance (X_s) and accounts for the effect of both armature reaction and leakage reactance. The equivalent circuit for an alternator then becomes as shown in Figure 5-6. The resistance is usually much smaller than the synchronous reactance and sometimes can be neglected.

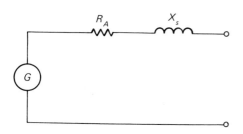

FIGURE 5-6   Equivalent circuit for an alternator

Transient Characteristics of Alternators

When a balanced short circuit is suddenly applied to an alternator, armature reaction tends to reduce the air gap flux. However, reducing the flux will induce currents in the rotor circuit that tend to maintain the flux; that is, it gives the effect of momentarily increasing the excitation. The result is that the short-circuit current will start off at a rather high value and gradually decay to a steady-state value, as shown in Figure 5-7.

The short-circuit current of an alternator has three time periods (see Figure 5-7). The first is the subtransient period, which lasts for about 5 or 6 cycles. The current is high due to induced currents in the damper winding and/or the rotor core and also currents induced in the excitation circuit. The next period is the transient period, which lasts for about 2 to 10 seconds. The induced currents in the damper windings have decayed to zero, but the currents induced in the dc rotor winding are still present. The third period is the steady-state period when all the induced rotor currents have disappeared, and the current that flows is due to the dc excitation applied.

Effects of Unbalanced Loading

As long as the load on an alternator is balanced, the effect of armature reaction (and therefore the air gap flux) is the same at all rotor positions. However, if the load is not balanced, some phases will create more mmf than others, and so the air gap flux will vary with rotor position. These variations of the flux will induce currents in the dc rotor winding, the damper winding, and also in the rotor core itself, and the result is overheating of the rotor.

Alternators are quite strictly limited as to the degree and duration of load unbalance that they can withstand. Current balance and/or negative sequence relays[1] are used to protect an alternator against this condition.

FIGURE 5-7   Short-circuit current from an alternator

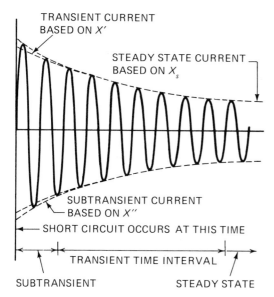

TRANSIENT CURRENT BASED ON X'

STEADY STATE CURRENT BASED ON X_s

SUBTRANSIENT CURRENT BASED ON X''

SHORT CIRCUIT OCCURS AT THIS TIME

TRANSIENT TIME INTERVAL

SUBTRANSIENT   STEADY STATE

---

[1]The topic of "symmetrical components" is beyond the scope of this book. For an introduction to the subject, see Donald Beeman, *Industrial Power Systems Handbook* (New York: McGraw-Hill Book Company, Inc., 1955).

Because there is considerable mutual inductance between the three phase windings of an alternator, balanced loading causes greater voltage regulation than single-phase loading. Whether line to neutral or line to line, single-phase loads cause unbalanced terminal voltages, and in the case of single-phase short circuits, the current flow will be higher than it is for three-phase faults.

Parallel Operation of Alternators

Because the demand for electrical energy continues to grow at a steady pace, electrical utilities (and others) find it necessary to increase generating capacity at regular intervals. Since it really is not economical to discard serviceable alternators in favor of larger ones, it becomes necessary to either operate alternators in parallel or else subdivide the load and feed it from multiple isolated systems. Of these two alternatives, parallel operation is much preferred because it provides greater reliability and does not require major changes to the existing system.

It is quite easy to start up an alternator and get it operating in parallel with existing units. This process is called "synchronizing the alternator," and we usually try to do it without disturbing the voltage or frequency of the existing system. However, before this is attempted we must be sure that the phase sequence of the incoming alternator matches that of the system to which it is going to be connected. The process of checking the phase sequence and getting it correct is known as "phasing out the alternator."

Figure 5-8 shows an alternator 1 supplying a load, with alternator 2 all wired up and ready for start-up. If these are low-voltage alternators, we can phase out alter-

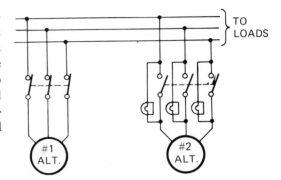

FIGURE 5-8   Parallel operation of alternators

nator 2 by connecting lamps across its disconnect switch, as shown in Figure 5-8. Start up machine 2, adjust its voltage and frequency to approximately the same values as the operating system, and watch the lamps closely. If the lamps flash on and off in unison, the phase sequence of unit 2 matches that of unit 1. If the lamps flash on and off sequentially, the phase sequence of unit 2 must be reversed relative to the system, and the usual way to do this is to interchange any two wires on either the generator side or the line side of the switch.

It is possible to achieve the same effect by reversing the rotation of the alternator, but this is very seldom convenient. Once alternator 2 has been properly phased out, it will always be correct unless either the wiring or the rotation is altered. We therefore go through the phasing out procedure only on the initial start-up of the unit.

With the three-lamp method of phasing out (shown in Figure 5-8), the voltage across each lamp can go as high as $(2/\sqrt{3})V_{LL}$. For 600-V alternators, that is about 694 V, and it becomes necessary to use many lamps in series across each pole of the switch. One can also phase out alternators using lamps across two poles of the switch and a solid jumper across the third (two-lamps-and-jumper technique). When correctly phased out, the lamps still flash in unison, but the voltage across each set of lamps will now reach a maximum of twice the line-to-line voltage.

For high-voltage alternators, direct connection of lamps is impractical. One could connect potential transformers directly across the switch and connect lamps into their secondary circuits. However, in most installations there will be potential transformers connected across the line on both sides of the disconnect switch. Phasing out can then be accomplished by connecting lamps as shown in Figure 5-9.

Lamps that are used to check phase sequence can also be used for synchronizing and so are often called synchronizing lamps.

The final act of synchronizing an alternator to an operating system is to close the disconnect switch that separates the two. However, we must meet four requirements before we close that switch:

■ The voltage of the incoming alternator

FIGURE 5-9 Phasing out using potential transformers and lamps

TO SYSTEM BUS

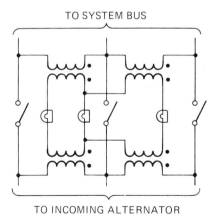

TO INCOMING ALTERNATOR

must match that of the system. Watch the switchboard voltmeters and adjust the alternator's excitation as necessary to meet this requirement. If the operator fails to do this, a voltage disturbance will be created on the system when the switch is closed.

■ The phase sequence of the incoming unit must match that of the system. If the alternator has been phased out as previously described, we need not give this any further thought, and for single-phase alternators this question does not apply.

■ The frequency of the incoming unit must match that of the system (or very nearly so). The rate at which the synchronizing lamps flash (flashes per second) is equal to the frequency difference in hertz. Adjust the governor setting on the alternator's prime mover as necessary to get the flash rate down to perhaps once in 5 seconds (or even slower). If this is not done, closing the switch will disturb both the voltage and the frequency of the operating system.

■ The voltages of the incoming machine must have the correct phase position relative to those of the system. Some people look at the loop (or loops) between the alternators, and from that viewpoint they say that "the voltages must be in phase opposition" and they are absolutely right. Other people look at the alternators from the position of the load and say that "the voltages must be in phase with each other." These apparently conflicting statements are really saying the same thing, but to make either statement without specifying the point of view leads to confusion.

However you describe the correct phase positions of the voltages, it is easy to see when the desired condition has been achieved. When the synchronizing lamps (connected for the three-lamp or two-lamp-and-jumper technique) are out, the desired position has been reached and the switch may be closed.

If the switch is closed when the alternator is out of phase with the system, a high circulating current will momentarily flow through the alternator and pull it ahead (or back) to its correct position. However, this creates a voltage disturbance on the system, and sometimes the magnetic forces generated on the alternator windings will damage the coils. For these reasons, it is important that the synchronizing operation be done as accurately as possible.

If the synchronizing operation has been correctly done, the alternator will now be "floating on the line"; that is, it will neither contribute to nor draw anything from the system. To make the alternator deliver true power (kilowatts), increase the driving torque applied to its shaft by its prime mover. The usual way of doing this is to raise the speed setting of the governor on the prime mover, but this tends to raise the frequency on the electrical system. The only way to prevent the frequency from changing is to lower the governor settings on the prime mover(s) of one or more other alternators on the system.

The reader should note that because of the rigid mathematical relationship between alternator speed and frequency there will be no change in the speed of any of the machines if this is being properly done. However, some redistribution of the reactive load will probably be observed.

The reactive power output (lagging volt-amperes reactive; VAr) delivered by the alternator can be increased by raising its field excitation. This tends to raise the system voltage, but we can prevent this by lowering the excitation on some other machine. The true

power load on the alternators is not affected by adjusting the excitation in this way.

To remove the alternator from the line, adjust the driving torque and excitation as necessary to bring its real and reactive power output to zero (adjust other machines as necessary to keep the voltage and frequency constant), and then open the switch. The prime mover can now be shut down if desired.

To permit more accurate synchronizing, most switchboards include a synchroscope that will indicate the relative phase position of the incoming alternator and also indicate whether the unit is running faster or slower than it should. It is usually easiest to watch the synchronizing lamps and to adjust the alternator speed to approximately the correct value. Then watch the synchroscope and do the final speed adjustment so that the incoming unit is running slightly fast. Once that has been done, continue to watch the synchroscope and close the alternator switch at (or just before) the time that the synchroscope reaches the zero position (zero indicates that the alternator is in the correct phase position).

The connection shown in Figure 7-8 (two-bright-and-one-dark method) is also sometimes used for synchronizing lamps. When the alternator is in the correct phase position, the lamp that is connected straight across the switch (center lamp in Figure 5-10) will be out, and the other two will be of equal brilliancy. This permits more accurate synchronizing than can be obtained using the "three-lamps-dark" technique and also has one other advantage. If the speed of the incoming alternator is incorrect, the lamps flash sequentially, and the order in which the lamps flash will tell the operator whether the incoming machine is running too fast or two slow. This makes it easier to adjust the alternator speed.

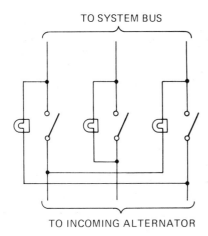

TO SYSTEM BUS

TO INCOMING ALTERNATOR

FIGURE 5-10   Two bright and one dark connection
of synchronizing lamps

There are two cautions to be observed with regard to synchronizing lamps. First, the lamp arrangement used on some switchboards will not check phase sequence. If for any reason an alternator is disconnected from the switchboard, normal operation of the lamps (when the alternator is reconnected) does not always guarantee that the phase sequence has not been inadvertently changed. Second, synchronizing lamps constitute a bypass around the alternator disconnect switch. In addition to locking the switch in the open position, the maintenance electrician must disconnect the synchronizing lamps to ensure his personal safety when working on the power circuit conductors.

SOME ASPECTS OF SYNCHRONOUS ALTERNATOR DESIGN

Hydrogen Cooling

If we attempt to use air to cool a large high-speed alternator, windage accounts for a substantial portion of the total losses. A considerable improvement can be obtained by designing the alternator with a gastight enclosure and filling it with hydrogen instead of air. Since hydrogen is a very light gas, windage losses are reduced, but at the same temperature and pressure, its heat-dissipating ability is about the same as air. To ensure that an explosive mixture will not accumulate within the alternator, the interior hydrogen is kept above atmospheric pressure, sometimes by as much as 400 kilopascals (kPa). The improved heat-carrying ability of pressurized hydrogen permits a higher output from a given physical size of machine.

Hydrogen cooling has two other advantages. It prevents deterioration of the insulation due to slow oxidation, and it actually reduces the fire hazard because the insulation cannot burn in the absence of oxygen.

As long as the hydrogen content exceeds 70%, an internal explosion cannot occur, and under normal conditions this is easy to maintain. However, before filling a new alternator with hydrogen, we must first purge the interior with carbon dioxide to drive out the oxygen and thereby prevent the occurrence of a potentially explosive condition.

Conductor Cooling

The materials used as electrical insulation on a machine winding tend to be fairly good thermal insulators as well. The heat produced by the copper losses in the machine is generated right in the conductors themselves and, with conventional cooling systems, has to flow through the insulation before it can be picked up by the hydrogen that circulates over the surface of the coils. The coil insulation therefore forms a thermal barrier that must somehow be overcome.

Normally, the conductors in a large alternator are not solid but instead are composed of many strands in parallel. If we replace a few of these strands with hollow tubes of the same external size, we can then circulate a fluid through those tubes and carry away the heat without having it flow through the coil-to-ground insulation. The result is substantially improved cooling, which permits a higher output from the same physical size of machine.

Conductor cooling (or inner cooling as it is sometimes called) may be applied to either or both the stator and the rotor. The fluid used for conductor cooling may be hydrogen, oil, or water.

MATHEMATICAL INTRODUCTION TO SYNCHRONOUS ALTERNATORS

Generated Voltage

Equation 4-22 is exact for synchronous alternators in which the field flux is sinusoidally distributed and a good approximation for a practical machine. Equations 2-4, 4-9, and 4-10 regarding pitch and distribution factors are also valid.

EXAMPLE 5-1

A 24-pole 60-Hz synchronous alternator has 432 coils in 432 slots, and the coil span is 15 slots. If each coil has 4 turns, the winding is 3 parallel wye, and the rotor produces 1.22 webers (Wb) per pole, find the line to line voltage.

Solution

$$\text{Full coil pitch} = \frac{432}{24} = 18 \text{ slots}$$

From Equation 2-4,

$$\text{Coil pitch} = 180 \times \left(\frac{15}{18}\right) = 150°$$

From Equation 4-9,

$$\text{Pitch factor } K_p = \sin\left(\frac{150}{2}\right)° = 0.96593$$

From Equation 4-10,

$$\text{Distribution factor } K_d = \frac{\sin[6(180/18)/2]}{6 \sin[(180/18)/2]}$$

$$= \frac{\sin 30°}{6 \sin 5°} = 0.95614$$

From Equation 4-1,

$$\text{Rotation speed (rev/s)} = \frac{2 \times 60}{24} = 5 \text{ rev/s}$$

From Equation 4-22,

$$\text{Phase voltage } |E_s| = \frac{1.11 \times 0.96593 \times 0.95614 \times (2 \times 4 \times 432) \times 1.22 \times 5}{3 \times 3}$$

$$= 2401 \text{ V}$$

$$\text{Line-to-line voltage} = 2401 \times \sqrt{3} = 4159 \text{ V} \qquad \text{(Answer)}$$

Voltage Regulation Calculations

On page 148 we explained how to construct a phasor diagram that can be used to calculate the voltage regulation of an alternator. It is essentially the same as that used to calculate voltage regulation of a transformer in Chapter 3, but the voltage drop due to reactance is of much greater magnitude in an alternator than it is in a transformer. We will confine ourselves to wye-connected three-phase alternators and will always assume a balanced load.

EXAMPLE 5-2

A 1000-kVA, 14,400-V three-phase alternator has 6.5-Ω effective resistance and 135-Ω reactance per phase. Assuming that it delivers rated voltage at full-load current, find its voltage regulation with

(a) A unity power factor load.
(b) An 80% power factor (lagging) load.
(c) A 60% power factor (leading) load.

Solution

Full-load current
$$= 1{,}000{,}000/(14{,}400 \times \sqrt{3}) = 40.09 \text{ A}$$

At full load,

$$I_A R_A = 40.09 \times 6.5$$
$$= 261 \text{ V}$$
$$I_A X_s = 40.09 \times 135$$
$$= 5412 \text{ V}$$

Rated voltage per phase $= \dfrac{14{,}400}{\sqrt{3}}$

$$= 8314 \text{ V}$$

(a) For a unity power factor load,

$$|E_s| = \sqrt{(8314 + 261)^2 + (5412)^2}$$
$$= 10{,}140 \text{ V}$$

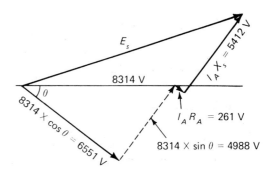

Voltage regulation

$$= \frac{10{,}140 - 8314}{8314} \times 100$$

$$= 21.96\% \qquad \text{(Answer)}$$

(b) For an 80% power factor (lagging) load, $\theta = \arccos 0.8 = 36.87°$.

From the phasor diagram,

$$|E_s| = \sqrt{(6551 + 261)^2 + (4988 + 5412)^2}$$
$$= 12{,}432 \text{ V}$$

Voltage regulation

$$= \frac{12{,}432 - 8314}{8314} \times 100$$

$$= 49.53\% \qquad \text{(Answer)}$$

(c) For a 60% leading power factor load, $\theta = \arccos 0.6 = 53.13°$.

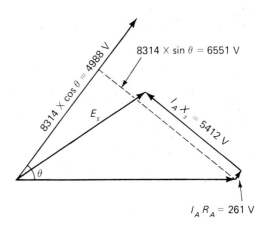

$$|E_s| = \sqrt{(4988 + 261)^2 + (6551 - 5412)^2}$$

$$= 5371 \text{ V}$$

Voltage regulation

$$= \frac{5371 - 8314}{8314} \times 100$$

$$= -35.4\% \qquad \text{(Answer)}$$

Synchronous Impedance Test

Looking at the equivalent circuit in Figure 5-6, it appears easy to determine the values of R_A and X_s by means of laboratory measurements. The procedure commonly used to do this is known as the synchronous impedance test; it is performed in two broad steps, more or less as follows.

Short circuit the stator windings through a suitable switch with ammeters arranged to measure the stator current. Drive the alternator at rated speed and raise the excitation until rated current is circulating in the stator windings. Holding the speed and excitation constant, remove the short circuit from the stator windings (i.e., open the switch), and measure and record the open-circuit voltage.

Next determine the terminal-to-terminal dc resistance of the stator winding using an ohmmeter or a voltmeter–ammeter technique.

If we now assume the stator winding to be wye connected, we can find ($|E_s|$), the generated voltage per phase (divide the observed open-circuit voltage by $\sqrt{3}$). The current that flowed while the stator was short circuited is the current per phase, and the synchronous impedance ($|Z_s|$) is just $|E_s|/|I|$. The resistance we measured is the dc resistance of two phases, so divide by 2 to get the dc resistance for one phase and multiply by 1.5 to get the effective ac resistance of that phase (R_A). We can now find the synchronous reactance ($|X_s|$) using Equation 5-1.

$$|X_s| = \sqrt{|Z_s|^2 - R_A^2} \qquad (5-1)$$

Once we have found the values of $|X_s|$ and R_A, we can calculate the voltage regulation of the alternator as outlined in the preceding section. However, if we actually load the alternator, we usually find that the regulation is quite a bit less than we calculated. There are two reasons for this. First, saturation of the iron tends to limit the no-load voltage and so tends to improve voltage regulation. Second, the nonuniform air gap of a salient pole machine causes it to possess lower $|X_s|$ when supplying a unity power factor load than it has when supplying purely reactive current. The mathematical techniques used to take these factors into account are basically extensions of the mathematics in the preceding section, but we will not go into them here.

Losses and Efficiency

The windage and friction losses are independent of load. The iron losses vary approximately as the square of the terminal voltage and so in most cases should be regarded as constant. For a given machine, the stator copper losses are dependent only on the load current, and the rotor copper losses are dependent upon the dc excitation current. Equation 1-13 applies.

EXAMPLE 5-3

Find the efficiency of the alternator in Example 5-2 at each load level. The rotor resistance is 2.7 Ω and the excitation current is 67, 86, and 41 A at full load and unity, 0.8 lagging, and 0.6 leading power factor, respectively. The windage and friction are 25 kW and the iron losses are 19 kW.

Solution

With a unity power factor load,

Rotor copper loss = $67^2 \times 2.7$ = 12.12 kW

Stator copper loss
$$= 3 \times 40.09^2 \times 6.5 = 31.34 \text{ kW}$$

Iron losses = 19. kW

Windage and friction = 25. kW

Total losses = 87.46 kW

$$\text{Efficiency} = \frac{1000}{1000 + 87.46} \times 100$$

$$= 91.96\% \qquad \text{(Answer)}$$

For a 0.8 power factor lagging load,

Rotor copper loss = $86^2 \times 2.7$ = 19.97 kW

Alternator output = 1000×0.8 = 800 kW

Efficiency
$$= \frac{800}{800 + 19.97 + 31.34 + 19 + 25} \times 100$$

$$= 89.35\% \qquad \text{(Answer)}$$

For a 0.6 leading power factor load,

Rotor copper loss = $41^2 \times 2.7$ = 4.538 kW

Alternator output = 1000×0.6 = 600 kW

Efficiency
$$= \frac{600}{600 + 4.538 + 31.34 + 19 + 25} \times 100$$

$$= 88.25\% \qquad \text{(Answer)}$$

Limits of Synchronous Alternator Loading

Most alternators are rated at some fixed kVA at a given power factor (usually 0.8 lag). At higher or at leading power factors, the kVA rating defines the maximum permissible load. However, at lower lagging power factors an alternator requires more excitation to maintain rated voltage, and so because of rotor heating it cannot deliver the rated kVA.

Let us consider a cylindrical rotor machine and neglect the effects of saturation and stator resistance. If the synchronous reactance is known and rated voltage and current are known, the load limits for the alternator can be described geometrically as follows (refer to Figure 5-11). Rated stator current can be allowed to flow for all leading power factor and high lagging power factor conditions down to the rated power factor of the machine. In Figure 5-11 this is shown by the circle (*AB*) whose center is at the origin.

Point *A* represents zero power factor leading, and *B* represents current at the rated power factor of the machine. Point *C* is just rated voltage per phase divided by synchronous reactance. Point *D* represents zero power factor lagging, and curve *BD* (which is a circle with center at *C*) defines the maximum load current at power factors below the rated value, that is, the maximum load that can be carried at rated voltage without exceeding the rated current in the rotor winding.

FIGURE 5-11 Limits of synchronous alternator loading

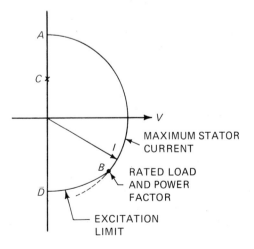

SUMMARY

The construction and theory of operation of an alternator are fairly simple. Armature reaction is a little more complex than that observed in dc machines but is quite similar in many ways. Phasing out, synchronizing, and parallel operation have some similarities to connecting and paralleling dc generators, and the differences can be understood in simple terms. Hydrogen cooling systems and conductor cooling systems are peculiar to alternators and have been important in the development of ever-larger machines.

Our mathematical treatment of alternators has contained very little new material. The generated voltage equation is common to both three-phase induction motors and synchronous motors, the voltage-regulation calculations are the same as those we used for transformers, and the losses and efficiency calculations are almost identical with those of other ac machines.

Only the synchronous impedance test and the limits of alternator loading are unique to this chapter.

REVIEW QUESTIONS

5-1. Explain what is meant by the following terms:
 (a) saturation curve
 (b) armature reaction
 (c) phase sequence
 (d) phasing out an alternator
 (e) synchronizing
 (f) floating on the line
 (g) synchroscope

5-2. What three features of an alternator are or may be different from those of a synchronous motor?

5-3. Why is the no-load saturation curve not a straight line?

5-4. How does the armature mmf affect the air gap flux in an alternator when the load has
 (a) Unity power factor?
 (b) Zero power factor lagging?
 (c) Zero power factor leading?

5-5. What three factors cause voltage regulation in an alternator and how are they accounted for in the equivalent circuit?

5-6. Why does the short-circuit current of an alternator start at a high value and gradually decay?

5-7. Give two reasons why unbalanced loads on an alternator are objectionable.

5-8. (a) What are the four conditions necessary for parallel operation of alternators?
 (b) Of the four conditions, which is usually checked only on the initial start-up and why?

5-9. Using diagrams as necessary, explain how to phase out two low-voltage alternators using lamps.

5-10. Draw a diagram showing how to phase out two high-voltage alternators.

5-11. Draw a diagram showing how to hook up the two bright and one dark arrangement of synchronizing lamps, and explain the two advantages of this arrangement.

5-12. Straighten out any misconceptions in the following statement: "If when synchronizing an alternator we close the switch with the alternator somewhat out of phase with the system, it stays that way. That is why synchronizing should be done as accurately as possible."

5-13. With regard to hydrogen cooling,
   (a) Why must the hydrogen be maintained at least slightly above atmospheric pressure?
   (b) What is the advantage of keeping the hydrogen substantially above atmospheric pressure?

   (c) What other advantages does hydrogen cooling offer?

5-14. Explain how conductor cooling is done and explain also why it is especially valuable in high-voltage machines that have thick layers of insulation on the coils.

5-15. Briefly explain how to do a synchronous impedance test.

5-16. Voltage-regulation calculations based upon a synchronous impedance test are usually somewhat in error. Explain why.

MATHEMATICS PROBLEMS

5-1. Find the synchronous speeds of the following alternators.
   (a) 8 pole, 60 Hz
   (b) 2 pole, 50 Hz
   (c) 30 pole, 60 Hz
   (d) 56 pole, 60 Hz

5-2. A 60-Hz alternator that has 600 single-turn coils, 600 slots, and 50 poles is connected 5 parallel wye. If the coil span is 9, find the flux per pole needed to produce a line voltage of 25 kV.

5-3. A 6-pole 50-Hz alternator has 144 coils in 144 slots. The coil span is 19, and each coil has 5 turns. The windings are connected 2 parallel wye, and the rotor produces 0.3 weber per pole. Find the generated voltage.

5-4. A 625-kVA alternator rated at 2400 V has 0.43-Ω resistance per phase and 5.2-Ω synchronous reactance per phase. Find its voltage regulation with the following:
   (a) A unity power factor load.
   (b) A 0.85 lagging power factor load.
   (c) A 0.9 leading power factor load.

5-5. Repeat problem 5-4 neglecting the stator resistance (R_A).

5-6. Assume that the alternator in problem 5-4 is rated at 0.8 power factor lagging. Find the maximum kVA load it can supply at the following:
   (a) 0.6 power factor leading.
   (b) 0.6 power factor lagging.
   (c) Zero power factor lagging.

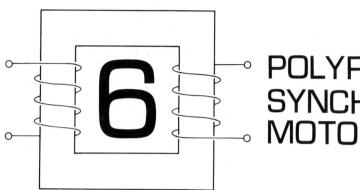

POLYPHASE SYNCHRONOUS MOTORS

In this chapter we will consider only the "doubly fed" type of three-phase synchronous motors, that is, the type that has three-phase stator windings and a dc rotor winding. From time to time synchronous motors that do not have any dc rotor winding but run strictly on reluctance torque have been produced, but they do not seem to enjoy much popularity, so they will be omitted. Single-phase synchronous motors will be dealt with in Chapter 7.

The conventional (doubly fed) type of synchronous motor has three virtues. First, it has zero speed regulation. For many applications, this is of no significance, but for some devices (like reciprocating compressors) it is a definite advantage. Second, it can operate at unity or even a leading power factor. This will not make any difference to the driven load, but it can reduce the cost of electrical energy (see the Appendix). Third, large low-speed synchronous motors are less expensive than similar motors of any other type.

We will start by describing the construction of a synchronous motor. Then we will go through its starting requirements and explain why it has its desirable characteristics. We will then examine running characteristics and quantify performance using an equivalent circuit.

Finally, we will mathematically treat its power factor correcting capabilities. Control equipment for synchronous motors is treated in Chapter 10.

CONSTRUCTION AND THEORY OF OPERATION

Synchronous Motor Construction

The construction of a synchronous motor is essentially the same as an alternator, but hydrogen cooling and conductor cooling systems are not commonly used. Ordinarily, there is no neutral connection to the stator windings, and almost without exception a synchronous motor will have damper windings that not only prevent hunting but are also used to start the motor. As with alternators, the stator is called the armature and the rotor is known as the field.

Starting a Synchronous Motor

When the stator windings are energized, the magnetic field they create rotates at a constant speed determined by the line frequency and the number of poles. The speed at which the field rotates is known as the synchronous speed of the motor and can be calculated using Equation 4–1.

The rotating field induces currents in the damper winding, and so torque is developed on the rotor, tending to move it along with the magnetic field. The motor will usually accelerate its load to about 95% of synchronous speed, running as a squirrel-cage motor in this fashion. If we now apply direct current to the insulated rotor winding, the rotor will pull into step with the rotating magnetic field (i.e., the rotor poles lock into step with the moving stator magnetic poles), and the motor is now said to be synchronized.

Starting a synchronous motor sounds simple and it is, but three points should be kept in mind.

■ The direct current to the rotor should be turned off during the starting period. Having the field winding energized with direct current cannot possibly contribute any average torque during acceleration, and in most motors it will actually reduce the torque. Keep the excitation off until the motor has reached its highest possible speed.

■ Most synchronous motors start on their damper windings. If the motor does not have damper windings, it will not be a self-starting design. Instead it will have to be driven at nearly synchronous speed by some external means and then synchronized by applying dc field excitation. In most motors the damper windings will overheat if the motor runs too long as a squirrel-cage machine. Synchronizing must be done promptly, within a minute or so of start-up.

■ The dc field winding must be shorted through an appropriate resistor during the starting period. The rotor winding is usually designed for either 125 or 250 V dc, but if left open during acceleration, the revolving magnetic field may induce an emf of several thousand volts in that winding, sufficient in many cases to break down its insulation. Short circuiting the winding protects it against this form of damage. One might be tempted to place a direct short circuit across the rotor winding during start-up, but it is not always practical to do so. The shorted dc rotor winding produces some torque during acceleration, but if you apply a direct short circuit to the rotor winding, you may give the motor a severe torque dip at half-speed, thereby preventing it from accelerating properly.

Mechanical Aspects of Synchronous Motor Operation

Assuming constant line frequency, the stator magnetic poles rotate at an absolutely constant speed. When the motor is synchronized, the rotor poles lock into step with the stator magnetic poles, but the connection between them is only magnetic lines of force and is therefore somewhat elastic. When we place a load on the motor, the rotor poles drop back a fraction of a turn behind the stator

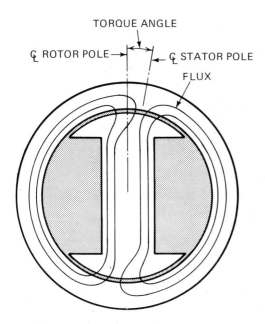

TORQUE ANGLE

℄ ROTOR POLE ← | → ℄ STATOR POLE

FLUX

FIGURE 6-1 Torque angle in a synchronous motor

poles; when the load is removed, the rotor catches up again, as shown in Figure 6-1. The angle between the rotor and stator poles is called the torque angle. Under no-load conditions, the torque angle is very small. Because the torque angle varies with load, the motor speed is not absolutely constant, but (as long as it remains synchronized) its average speed will be constant. A synchronous motor is therefore described as an average constant-speed motor.

Rather than work with mechanical degrees, let us specify torque angle in electrical space degrees, using the definitions from Chapter 2 with positive values indicating that rotor poles are lagging stator poles.[1]

If we suddenly apply a load to a synchronous motor, the torque angle cannot instantaneously change. Instead, the rotor

[1]Equation 2–4 can be applied to the stator if desired.

more or less gradually swings back and, because of inertia, the torque angle will momentarily exceed the required value. Usually the torque angle will vary up and down a few times before settling down to its new required value. This momentary oscillation of the rotor (relative to the stator poles) is known as hunting. Hunting induces momentary currents in the damper winding, and this tends to choke out (or damp out) the oscillation; but without the damper winding, hunting can become continuous and intolerably severe, sometimes enough to cause loss of synchronism.

Neglecting reluctance torque (which is dealt with later), the torque of a synchronous motor is proportional to the sine of the torque angle. Maximum torque is therefore developed at a torque angle of 90° electrical (half a pole width). If the motor is loaded beyond that point, it will pull out of synchronism and stall (it may continue to run as a squirrel-cage motor). The minimum torque load that will cause the motor to lose synchronism is called the pull-out torque of the motor. Reducing either the ac voltage applied to the stator or the dc excitation applied to the rotor will decrease the pull-out torque.

Pull-in torque is the maximum torque load with which the motor will synchronize. Low ac voltage on the stator or low dc excitation will reduce pull-in torque, but there are two other factors that affect it. High load inertia makes synchronizing more difficult and therefore reduces pull-in torque. The remaining factor is the relative position of the rotor and stator poles at the instant that direct current is supplied to the rotor. This last point introduces complications to the control equipment used with synchronous motors and is dealt with in Chapter 10.

Electrical Aspects of Synchronous Motor Operation

The revolving magnetic field in either synchronous or induction motors generates a counter emf in the stator windings that limits the current flow there. However, it must be remembered that the stator windings create the revolving field by drawing a current that lags the applied voltage by 90°. If the torque angle in a synchronous motor is zero (it is almost so at no load), the rotor and stator mmf's are coincident and additive. It therefore follows that raising the dc excitation will reduce the magnetizing current required by a stator winding. If we adjust the excitation to an appropriate level, the magnetizing current flow in the stator goes to zero. If we raise the excitation above that point, the stator begins to draw a leading current. The effect of this leading current is to oppose the rotor mmf, and the rotating magnetic field (i.e., the air gap flux) remains almost constant. Even with a load on the motor, changing the excitation has essentially the same effect.

At any load, normal excitation is that which reduces the stator magnetizing current to zero. At any given load the real power input (watts) to the stator remains constant, and so normal excitation is also that which results in minimum stator current.

Increasing the torque load on the motor will cause an increase of stator current and true power input with only a modest increase of stator magnetizing current. But changing the excitation has almost no effect on true power input, because the change of apparent power is accompanied by an inverse change of power factor.

A graph that shows stator current (ordinate) at various values of dc excitation (abscissa) is known as a V-curve. Figure 6-2

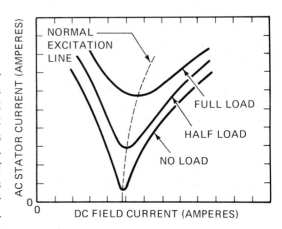

FIGURE 6-2   Typical V curves for a synchronous motor

shows three V-curves for a typical synchronous motor at full-, half-, and no-load, respectively. As the figure implies, increasing the load on the motor raises the V curve, rounds out the bottom of the V, and shifts the curve to the right.

It is convenient at this point to introduce a simple equivalent circuit for a synchronous motor and illustrate the effects of changing mechanical load and/or excitation using phasors. If we were to apply field excitation to a synchronous motor and drive it with a gasoline engine, the rotating magnetic field would generate a voltage in the stator windings. This same generating action occurs whenever the motor is running. So it is valid to talk about the voltage generated in the stator due to the magnetic field of the rotor, even though we cannot readily measure that voltage when the motor is connected to the supply line (you can measure it if you suddenly switch off the supply lines, leaving the motor coasting). We will represent each phase of a synchronous motor as a rather high impedance (mostly reactance) connected in series with an internal voltage source, as shown in Figure 5-6. We will use the symbol

E to denote this internally generated voltage (which exists because of rotor excitation) and V to indicate the applied voltage. We will not insert double subscripts but will always assume the same subscripts in the same order for both V and $E;$ and since all three phases will be the same, we will draw phasors for only one. When required, we will use the subscript A to represent armature (i.e., stator) quantities so that Z_A, R_A, and X_A represent stator winding impedance, resistance, and reactance, respectively.

Figure 6-3 shows what happens when the excitation is adjusted so that $|E|$ equals $|V|$ and load is applied to the motor. The angle α is always equal to the torque angle, and as load is applied, E swings clockwise as shown. The phasor sum of E and V (identified as E_R) therefore follows a circular locus; and since $I_A = E_R/Z_A$, the stator current will also follow a circular path in such a fashion as to remain almost 90° behind E_R. The power input to the motor (per phase) is just $VI_A \cos \theta$, and neglecting copper losses due to stator winding resistance, torque

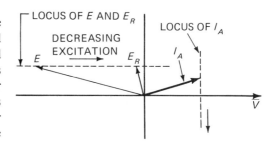

FIGURE 6-4   Effect of changing excitation on a synchronous motor

is proportional to the power input. Maximum torque is therefore developed when E_A is at 90°, which puts I_A about 45° behind V.

Figure 6-4 shows the effect of changing the excitation on the motor with a constant torque load. As the dc field current is increased, E travels along a straight line to the left (i.e., the torque angle decreases), E_R swings in the counterclockwise direction (changing its magnitude so as to follow a straight line), and I_A follows a vertical line, remaining 90° behind E_R. The true power input remains unchanged, but the reactive power ($VI_A \sin \theta$) changes considerably. As excitation increases, I_A (and therefore the power factor) swings in the leading direction. If excitation is sufficiently reduced, E will retract and swing to the 90° position and the motor will lose synchronism.

The effect of changing load on an overexcited synchronous motor is shown in Figure 6-5. Note that with no load the stator current leads the applied voltage by very nearly 90°. A motor designed to operate in this condition is known as a synchronous capacitor and can be used to correct a lagging power factor condition caused by other loads on the system. But most synchronous motors are designed to drive some specified mechanical load when operated at either 1.0 or 0.8 power factor leading.

FIGURE 6-3   Effect of load torque on a synchronous motor

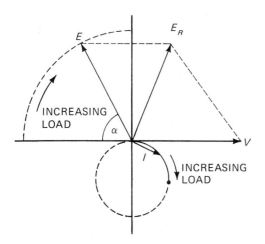

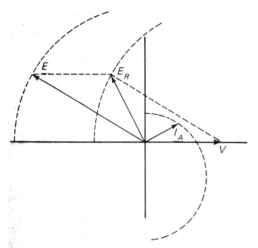

FIGURE 6-5   Effect of changing load on a
overexcited synchronous motor

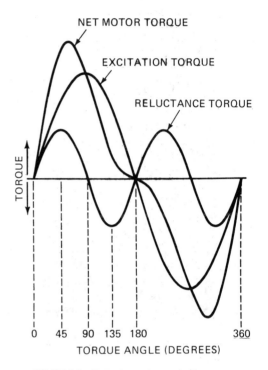

FIGURE 6-6   Excitation torque and reluctance
torque

Reluctance Torque

Note that rotors like those shown in Figure 5-1 result in a nonuniform air gap between the rotor and stator. Because they provide a lower reluctance magnetic path, the projecting poles of the rotor tend to lock into step with the stator poles, even with no excitation. This reluctance torque (as it is called) is proportional to the sine of twice the torque angle and independent of excitation. Reluctance torque has two interesting effects. First it raises the pull-out torque of the motor and decreases the angle at which it occurs, as can be seen in Figure 6-6. Second, with light or no load, reluctance torque may be enough to synchronize the motor before excitation is applied. If this happens, it may synchronize with a torque angle of either about 30° or 210°; in the latter case, the motor will be forced to slip a pole when sufficient excitation has been applied. The stator current and power factor will swing rather wildly when this happens, but once the motor has slipped to the proper torque angle, it will operate normally.

SOME ASPECTS OF SYNCHRONOUS MOTOR DESIGN

Stator Core and Winding Assembly

The stator core and windings of a synchronous motor are essentially the same as those of a squirrel-cage machine, but they have a few special refinements. First, since rotor excitation can reduce or even eliminate any magnetizing current in the stator, it is quite practical to build these machines with large air gaps and for low-speed operation. Synchronous motors frequently have 30 or 40 or even more poles, whereas induction motors rarely have more than 14. Second, there are few synchronous motors smaller than 50 kW per rev/s (i.e., full-load torque of less than 8000 newton-meters). Machines of this size nearly always have form-wound (rather than random or mush wound) coils. Third, multi-speed synchronous motors are almost non-existent because of the difficulties associated with a two-speed rotor design.

Rotor Designs

The rotor must always have the same number of poles as the stator with which it is to be used. However, its overall shape depends upon the speed at which it runs. For a low-speed motor (which has a large number of poles), the rotor may have a diameter of three to five times the axial length of the magnetic core; in a large high-speed machine the rotor core length may be three or four times the diameter.

The rotor core for a large high-speed motor is generally a solid-steel forging. Slots are machined into the rotor periphery (parallel to the shaft), and the dc field winding is placed in these slots. Figure 6-7 shows

the typical cross section for large two- and four-pole rotors. The two-pole rotor shown has parallel slots, but radial slots (like those of the four-pole rotor) have also been used. Because of their overall shape, these are known as cylindrical rotors, and the advantage of this construction is that it withstands the very high centrifugal forces that exist in large high-speed machines. These rotors develop only a small reluctance torque.

Smaller, lower-speed synchronous motors usually have salient[2] pole rotors. The pole cores are usually laminated and either dove-tailed to a steel hub, as shown in Figure 6-8, or (for low-speed machines) simply bolted to a cast-iron spider, as shown in Figure 6-9. Actually, each pole core has an insulated coil and a damper winding installed on it before it is mounted on the hub or spider. The segments of the damper winding are then bolted together and the insulated coils connected in series with each other. All that remains to be done is to provide some way of feeding direct current through this field winding.

Salient pole rotors develop considerable

[2]The word salient means "projecting."

FIGURE 6-7   Cylindrical cross sectons (two- and four-pole)

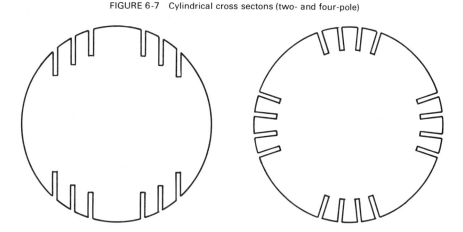

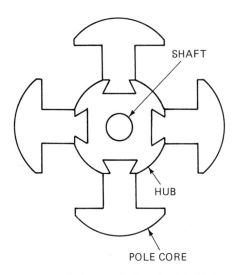

FIGURE 6-8   Salient pole cores dovetailed to a
central hub

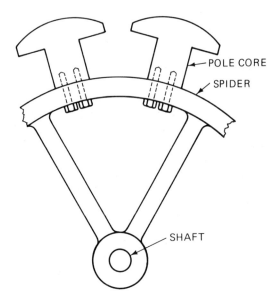

FIGURE 6-9   Salient pole cores bolted to a spider

reluctance torque, sometime enough to pull more than half their rated load.

Excitation Arrangements

The dc field windings of synchronous motors have traditionally been connected to a pair of slip rings. Stationary carbon brushes can then provide the electrical connections be-

tween the rotor winding and any external dc supply.

Historically, the shunt (or compound) wound dc generator has been used to excite (i.e., feed power to) the synchronous motor field. Such a generator may be either belt driven or mounted directly on the motor

FIGURE 6-10   Brushless synchronous motor

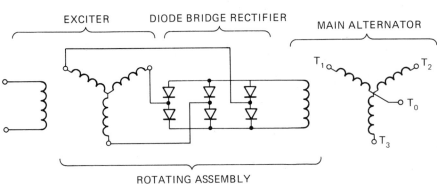

shaft, and sometimes a separate motor-generator set is used. There are other possibilities, such as the use of a rectifier assembly to provide direct current, but all these schemes use the slip-rings-and-brushes arrangement, which is a source of maintenance problems. For this reason, brushless synchronous motors have been developed. Figure 6-10 shows a schematic diagram for a brushless synchronous motor.

MATHEMATICAL INTRODUCTION TO SYNCHRONOUS MOTORS

Voltage and Flux Relationships

Equation 4–22, which shows the relationship between voltage and flux in induction motors, applies equally well to synchronous machines with sinusoidally distributed flux. However, if the flux is not sinusoidally distributed along the air gap (it usually deviates considerably therefrom in a salient pole machine), Equation 4–22 becomes only a reasonable approximation. Pitch and distribution factors are calculated using Equations 2–4, 4–9, and 4–10, as required.

EXAMPLE 6-1

A 10-pole, three-phase synchronous motor has 240 stator coils (total) and runs at 12 rev/s. The coil span is 19 slots, the applied voltage is 2300 V, and the stator coils are connected to form a series wye conventional winding. If each coil has 8 turns of wire, find the flux per pole in the magnetic circuit.

Solution

Using Equation 2–4,

$$\text{Coil pitch} = 180 \times \frac{19}{24} = 142.5°$$

Therefore, using Equation 4–9,

$$K_p = \sin\left(\frac{142.5}{2}\right)° = 0.94693$$

Using Equation 4–10,

$$K_d = \frac{\sin[8(180/24)/2]}{8 \sin[(180/24)/2]}$$

$$= \frac{\sin 30°}{8 \sin 3.75°}$$

$$= 0.9556$$

Using Equation 4–22,

$$\frac{2300}{\sqrt{3}} = 1.11(0.94693)(0.9556) \times (2 \times 8 \times 80)\phi(12)$$

$$\phi = 0.08607 \text{ Wb/pole} \quad \text{(Answer)}$$

Torque, Torque Angle, and Power in Cylindrical Rotor Machines

For these calculations we will use the equivalent circuit in Figure 5-6 and phasor diagrams similar to those in Figures 6-3 through 6-5. The symbols here will be the same as used on pp. 164–65, and unless otherwise noted all quantities are given on a per phase basis. The effects of a nonuniform air gap have been neglected.

There are three important angles in the phasor diagram in Figure 6-5. First there is α, the torque angle, which is considered to be positive when E_A is in the second quadrant. The second is θ, the angle between V and I_A, and it will be considered positive when I_A leads V. The third angle is β, the angle between E and I_A, and it will be considered positive when E leads I_A. The sum of the three angles is always 180°.

The four essential equations for synchronous motor torque angle and power calculations are as follows:

$$E_R = E + V \qquad (6\text{-}1)$$

$$I_A = \frac{E_R}{Z_A} \qquad (6\text{-}2)$$

$$P \text{ (input)} = VI_A \cos\theta \qquad (6\text{-}3)$$

$$P \text{ (developed)} = EI_A(-\cos\beta) \qquad (6\text{-}4)$$

The difference between the power input and power developed is the stator copper loss, which could also be calculated from Equation 6-5.

$$\text{Stator copper loss} = I_A^2 R_A \qquad (6\text{-}5)$$

Equation 2-3 applies to the torque and power developed by a synchronous motor.

The resistance of the armature winding in a synchronous motor is much smaller than the reactance, and in many cases R_A can be neglected without serious loss of accuracy.

If $R_A = 0$, then $Z_A = X_A$, and the stator copper losses become zero. Under these conditions, power developed becomes equal to the input power, which can be calculated using Equation 6-6.

$$P \text{ (input or developed)} = |EV|\sin\alpha \qquad (6\text{-}6)$$

EXAMPLE 6-2

A wye-connected synchronous motor rated at 2300 V line to line runs at 7.5 rev/s. The stator reactance is 1.3 Ω/phase and its resistance is 0.08 Ω/phase. Draw a phasor diagram and find the current, power factor, power input, copper losses, and torque output when the excitation is adjusted to make $E = 1600$ V/phase and the torque angle is 40°. Assume that windage and friction losses are zero.

Solution

$$E_R = E + V$$

$$= 1600\ \underline{/140} + \frac{2300}{\sqrt{3}}\ \underline{/0}$$

$$= 1600\ \underline{/140°} + 1328\ \underline{/0}$$

$$= -1225.7 + j1028.5 + 1328$$

$$= 102.33 + j1028.5$$

$$= 1033.5\ \underline{/84.32°}$$

$$Z_A = \sqrt{1.3^2 + 0.08^2}\ \arctan\left(\frac{1.3}{0.08}\right)$$

$$= 1.3025\ \underline{/86.48°}$$

$$I_A = \frac{E_R}{Z_A}$$

$$= 793.5\ \underline{/-2.16°} \qquad \text{(Answer)}$$

Power factor

$$= \cos(-2.16°)$$

$$= 0.99929 \text{ lagging} \qquad \text{(Answer)}$$

P (input)

$$= 1328 \times 793.5 \times 0.99929$$

$$= 1{,}053{,}020 \text{ W/phase} \qquad \text{(Answer)}$$

Copper loss

$$= I_A^2 R_A$$

$$= 793.5^2 \times 0.08$$

$$= 50{,}371.4 \text{ W} \qquad \text{(Answer)}$$

P (developed)

$$= 1600 \times 793.5 \times (-\cos 143.16)$$

$$= 1{,}002{,}637 \text{ W} \qquad \text{(Answer)}$$

$$\text{Torque} = \frac{1{,}002{,}637}{2 \times 7.5}$$

$$= 21{,}277 \text{ N} \cdot \text{m} \qquad \text{(Answer)}$$

EXAMPLE 6-3

With the excitation set to make $|E| = 3000$ V/phase, a wye-connected synchronous motor rated at 4000 V line to line draws 368 A at no load. If the stator winding resistance is 0.2 Ω/phase, and windage and friction are zero, draw appropriate phasor diagrams and find the following:

(a) The no-load copper losses.
(b) The no-load torque angle.

(c) Stator reactance (X_A).
(d) Power input, power output, and copper losses at a torque angle of 30°.

Solution

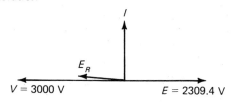

(a) $P = 368^2 \times 0.02 = 2708$ W.
(b) Apparent power per phase $= 368 \times 4000/\sqrt{3} = 849{,}890$ VA.

$$\therefore Pf = \frac{2708}{849{,}860} = 0.0031864$$

$$\therefore \theta = 89.82°$$

Since the developed power is zero, $\beta = 90°$.

$$\therefore \alpha = 180 - (90 + 89.82)$$

$$= 0.18° \quad \text{(Answer)}$$

(c) $E_R = 3000 \ \underline{/179.82°} + \left(\dfrac{4000}{\sqrt{3}}\right) \underline{/0}$

$$= -2999.99 + j9.42 + 2309.4$$

$$= -690.59 + j9.42$$

$$= 690.65 \ \underline{/179.22°}$$

$$Z_A = \frac{E_R}{I_A} = \frac{690.65 \ \underline{/179.22}}{368 \ \underline{/89.82°}}$$

$$= 1.8768 \ \underline{/89.4°}$$

$$X_A = 1.8768 \sin 89.4°$$

$$= 1.8767 \ \Omega \quad \text{(Answer)}$$

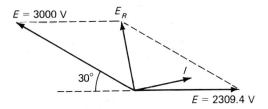

(d) $E_R = 3000 \ \underline{/150} + 2309.4 \ \underline{/0°}$

$$= -2598.1 + j1500 + 2309.4$$

$$= -288.7 + j1500$$

$$= 1527.5 \ \underline{/100.9°} \text{ V}$$

$$I_A = \frac{1527.5 \ \underline{/100.9°}}{1.8768 \ \underline{/89.4°}}$$

$$= 813.9 \ \underline{/11.5°}$$

P (input)
$$= 2309.4 \times 813.9 \times \cos 11.5°$$
$$= 1{,}841{,}887 \text{ W/phase} \quad \text{(Answer)}$$

P (developed)
$$= 3000 \times 813.9 \times (-\cos 138.5°)$$
$$= 1{,}828{,}725 \text{ W/phase} \quad \text{(Answer)}$$

Copper losses
$$= 813.9^2 \times 0.02$$
$$= 13{,}249 \text{ W} \quad \text{(Answer)}$$

Losses and Efficiency

Use Equation 1–14 to calculate efficiency. The losses in a synchronous motor are as follows:

Copper Losses

STATOR COPPER LOSS These losses are proportional to the square of the ac line current flowing to the motor. Increasing the mechanical load or operating at reduced power factor increases the stator copper loss. In effect, the heat produced by these losses establishes the maximum permissible continuous current rating of the stator winding.

If it becomes necessary to determine the resistance of the stator winding, one immediately thinks of measuring it with an ohmmeter or a voltmeter–ammeter technique. However, the results obtained by these methods will be the resistance to the flow of direct current, and the effective ac resistance will be somewhat higher. In the absence of any better information, assume that the effective ac resistance is 50% higher than the dc resistance.

ROTOR COPPER LOSS These are proportional to the square of the dc excitation current and in that sense are independent of load. However, if you raise the mechanical load on the motor and try to hold the power factor constant (or if you have the load constant and try to operate at a more leading power factor), you need more excitation current, resulting in increased rotor heating.

Iron Losses

These are just hysteresis and eddy current losses in the stator core and are essentially constant, regardless of load or excitation.

Friction and Windage

These losses are constant regardless of load or excitation. Treat them as a mechanical load on the motor.

EXAMPLE 6-4

A given synchronous motor has the following ratings and/or losses.

$$|V| = 2400 \text{ V}, \quad |I_{fL}| = 450 \text{ A}, \quad pf = 0.8$$
$$R_A = 0.16 \text{ } \Omega/\text{phrase},$$

Windage and friction = 19.5 kW

Iron loss = 34.6 kW

Find the following:

(a) The power input at no load if the line current is 25A.
(b) The power output at full load.
(c) The full-load efficiency.

Solution

(a) No-load copper loss = $3 \times 25^2 \times 0.16 = 300$ W.

∴ No-load power input
   = 19,500 + 34,600 + 300
   = 54,400 W   (Answer)

(b) At full load, copper loss = $3 \times 450^2 \times 0.16 = 97,200$ W.

∴ Total losses
   = 97,200 + 19,500 + 34,600
   = 151,300 W

Input = $\sqrt{3} \times 2400 \times 450 \times 0.8$
   = 1,496,492 W

∴Output = 1,496,492 − 151,300
   = 1,345,192 W   (Answer)

(c) Efficiency = $\dfrac{\text{output}}{\text{input}}$

   = $\dfrac{1,345,192}{1,496,492}$

   = 0.8989   (Answer)

Limits of Synchronous Motor Performance

Consider a synchronous motor operating at rated voltage. At or above rated power factor, stator copper loss determines the maximum input kVA. Below rated power factor (in the leading direction), rotor heating sets the excitation limit, and this defines the maximum permissible input to the stator.

FIGURE 6-11   Limits of synchronous motor performance

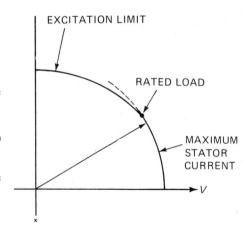

The result is that the maximum stator input is limited as shown in Figure 6-11.

The mathematical arguments are essentially the same as those for synchronous alternators (Chapter 5), and so they are not included here. The only likely point of confusion is the use of the terms "leading" and "lagging" with respect to power factor. Because the real power flow is generally into a motor, we take the power factor as the cosine of the angle between V and I with positive current flow defined to be into the motor. On an alternator, the power factor is still the cosine of the angle between V and I, but positive current is now taken to be out of the machine, and this results in the reversed terminology.

Power Factor Correction

It has already been stated that a synchronous motor can be used to correct the power factor of a given installation, but it is useful to be able to calculate the improvement that will be obtained. Two things about industrial practice should be understood. First, almost all industrial plants inherently operate at a lagging power factor. Second, synchronous motors are rarely (if ever!) operated at a lagging power factor; most of them operate at 1.0 or 0.8 lead or somewhere in between.

In doing any mathematics, remember that load watts (or kilowatts) add, while leading and lagging reactive (VAr) loads cancel or subtract. Here are two sample problems.

EXAMPLE 6-5

An existing plant has a total load of 9000 kVA at 72% power factor lag. The management decides to add a compressor driven by a 1000-kW synchronous motor. Assuming 90% efficiency, find the new overall power factor of

the plant if the motor is (a) a unity power factor type, and (b) a 0.8 power factor design.

Solution

Existing true power load

$$= 9000 \times 0.72$$
$$= 6480 \text{ kW}$$

Existing reactive power load

$$= 9000 \times \sin (\arccos 0.72)$$
$$= 6246 \text{ kVAr}$$

True power input to new motor

$$= 1000 \div 0.9$$
$$= 1111 \text{ kW}$$

(a) If the motor has 1.0 pf, its VAr input is zero:

New total load $= 6480 + 1111$
$$= 7591 \text{ kW and}$$
$$6246 \text{ kVAr}$$

Apparent power $= 7591^2 + 6236^2$
$$= 9830 \text{ kVA}$$

$$\text{Overall pf} = \frac{7591}{9830}$$
$$= 0.7722 \quad \text{(Answer)}$$

(b) If the motor runs at 0.8 pf lead,

Motor reactive power input

$$= 1111 \times \tan (\arccos 0.8)$$
$$= 833.25 \text{ kVAr}$$

The new total load is 7591 kW and $6246 - 833.25 = 5412.75$ kVAr.

New apparent power

$$= \sqrt{7591^2 + 5412.75^2}$$
$$= 9323 \text{ kVA}$$

$$\text{New power factor} = \frac{7591}{9323}$$
$$= 0.8142 \quad \text{(Answer)}$$

EXAMPLE 6-6

Referring to Example 6-5, at what power factor would the motor have to operate in order to get an overall power factor of 0.9?

Solution

From the work done in solving the previous example, we know that the new overall load will be 7591 kW, and since the new power factor is to be 0.9, the net reactive load must be

7591 × tan (arcos 0.9) = 3676.5 kVAr

Since the original load was 6246 kVAr, the reactive power input to the motor must be

6246 – 3676.5 = 2569.5 kVAr

Since the true power input to the motor is 1111 kW, the motor power factor is

$$\cos\left[\arctan\left(\frac{2569.5}{1111}\right)\right] = 0.3969 \quad \text{(Answer)}$$

SUMMARY

The synchronous motor is not a very complicated machine. Its stator is much like a three-phase motor; only the rotor is different. Its operation is readily understood in qualitative physical terms, and it is easy to see why it has high (or leading) power factor, constant average speed and is suitable for low-speed designs. The scarcity of small synchronous motors is mainly due to their cost. The rotor is inherently more expensive than a squirrel-cage type, and the low production volume also pushes the price of the synchronous motor well above that of a squirrel-cage machine.

A synchronous motor can be represented by a rather simple equivalent circuit. Using that circuit, it is easy to calculate the effect of changing the load or the excitation on the torque angle, the line current, and the power input.

Calculating the effect a synchronous motor has on the overall power factor of the supply system is just a matter of combining (or adding) loads that operate at different power factors. This type of question is usually found in electrical theory textbooks and is included here only because it arises in conjunction with synchronous motors.

REVIEW QUESTIONS

6-1. Explain what is meant by the following terms:
    (a) stator
    (b) rotor
    (c) field
    (d) armature
    (e) damper winding
    (f) synchronized
    (g) excitation
    (h) torque angle
    (i) hunting
    (j) air gap
    (k) normal (excitation)
    (l) V-curve
    (m) synchronous capacitor
    (n) reluctance torque
    (o) salient pole
    (p) cylindrical rotor
    (q) brushless synchronous motors
    (r) pull-in torque
    (s) pull-out torque

6-2. When starting a synchronous motor,
    (a) How does the motor develop torque?
    (b) Why should the excitation be removed?
    (c) Why should the field winding be short circuited through a resistor?

(d) Why should excitation be applied as soon as the motor has reached its highest possible speed?

6-3. "A synchronous motor has zero speed regulation but its speed is not absolutely constant." Explain.

6-4. What feature of a synchronous motor reduces hunting and how does it work?

6-5. What three factors affect (a) pull-out torque, and (b) pull-in torque?

6-6. Why doesn't a synchronous motor necessarily draw a magnetizing current from the line?

6-7. Assuming a constant load, how can the power factor of the motor be reduced in the leading direction?

6-8. If a synchronous motor is to be manually controlled, it is almost essential to have a stator ammeter in the circuit. Why?

6-9. Can the operator change the true power input to a synchronous motor? Explain.

6-10. Draw a phasor diagram to show what happens when the following occurs:
(a) Load is removed from a normally excited synchronous motor.
(b) Excitation is reduced on a normally excited synchronous motor.

6-11. An electrician moved the field rheostat slightly on a synchronous motor controller, in such a way as to increase the resistance. As he did so the stator current decreased. Was the motor running above or below normal excitation? Explain.

6-12. A student started a small synchronous motor with no load and gingerly applied a little bit of dc excitation. As he did so, he noticed that the line current went up instead of down. If the equipment is all in good working order, (a) what is wrong? (b) How can the problem be corrected?

6-13. Name one advantage of salient pole rotor construction.

6-14. Why is the synchronous motor better suited to low-speed designs than a squirrel-cage machine?

6-15. Draw a schematic diagram of a brushless synchronous motor and explain how it works.

6-16. What are the losses of a synchronous motor?

6-17. Explain how to get an approximate value for the effective ac resistance of a synchronous motor stator.

6-18. Unity power factor synchronous motors tend to be smaller and more efficient than 0.8 power factor types. Why?

6-19. Explain how a synchronous motor can improve the overall power factor of a plant.

MATHEMATICS PROBLEMS

6-1. Make a table showing full-load speeds for all synchronous motors operated from 50- and 60-Hz systems, 16 to 40 poles inclusive.

6-2. A 30-pole, 60-Hz, 4160-V synchronous motor is to be designed so as to have about 0.3 Wb/pole. The stator has 360 slots, the winding is conventional series wye, and the coil pitch is 9 slots. How many turns of wire will be required in each coil? Give a practical answer.

6-3. A 6-pole 50-Hz synchronous motor has 144 slots in the stator. The coil span is 19, and the winding is a conventional type connected three parallel delta. What applied voltage will produce 1.4 Wb total air gap flux?

6-4. A 13.2-kV motor has 28-Ω reactance per phase in its stator windings, which are connected series wye. Neglect windage, friction, and stator winding resistance. If the excitation is adjusted so that the internally generated voltage will be 9 kV, find the following:

(a) The current at a torque angle of zero.

(b) The current and power output at a torque angle of 28°.

(c) The maximum power output and the line current under that condition.

6-5. Assume that the motor from problem 6-4 has 0.5-Ω resistance per phase. Find the following:

(a) The no-load torque angle.

(b) The output when the torque angle is 28°.

(c) The copper losses at a 28° torque angle.

6-6. Assume that, in addition, the motor in problem 6-5 has 25-kW iron loss and 40 kW of windage and friction. Find the following:

(a) The no-load torque angle.

(b) The output when the torque angle is 28°.

(c) The efficiency at a 28° torque angle.

6-7. Raise the internally generated voltage to 10 kV and repeat problems 6-4 through 6-6.

6-8. Find the minimum no-load line current for the motor in problem 6-6, and find the value of E that results in the minimum no-load current.

6-9. An existing plant has a peak load of 6000 kVA at 0.8 power factor. To increase production, they are going to add a 750-kW synchronous motor. If the motor is 88% efficient, find the following:

(a) The overall power factor if a unity pf motor is installed.

(b) The overall power factor if a 0.8 pf motor is used.

(c) The power factor at which the motor must operate to obtain an overall pf of (i) unity, (ii) 90%, and (iii) 85%.

6-10. Referring to problem 6-9, and assuming no change of efficiency, how many kilowatts (output rating) of 0.8 power factor synchronous motors would be required to obtain an overall power factor of (a) 100%; (b) 90%.

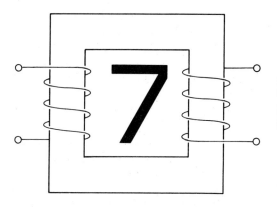

SINGLE-PHASE MOTORS

There are two basic reasons for the use of single phase rather than three-phase motors. First, there is the economics of the motor and its branch circuit. Fixed loads requiring not more than 0.5 kW can generally be served most economically with single-phase power and a single-phase motor.

To avoid the need of special branch circuits, portable tools are also usually single phase.

The second factor is the economics of power distribution. Where one live wire and an earth return path will suffice (such as electric railroads, oil fields, and farming communities), single-phase power distribution is often more economical than three phase. Homes and small commercial and industrial buildings usually have three-wire single-phase wiring systems, and the utility company will feed a whole group of such buildings from a suitable transformer energized from only one phase of the primary line. For these consumers, three phase is simply not available, at least not at an acceptable price.

Electric railroads use motors on the order of 1000-kW output, but most single-phase motors are rather small, ranging from about 10 W to about 1 kW. There are millions of these small motors in use, and the competition among the suppliers is very keen. Although many different varieties of single-phase motors have been produced over the years, competition has effectively reduced this array to the squirrel-cage types, the synchronous motors, and the universal motor.

Because they present the most challenges to trades people, and also because their mathematics is similar to three-phase induction motors, most of this chapter is devoted to squirrel-cage types. Their construction and theory of operation will be discussed, and it is assumed that the reader is familiar with pertinent material in Chapter 4. Schematic diagrams and brief descriptions of most commercial squirrel-cage single-phase motors are presented in this chapter. For readers interested in some of the basics of design, an abbreviated mathematical treatment is included. However, the reader should know the mathematics presented in Chapter 4 before attempting this material.

For those who require a basic understanding of them, other types of single-phase motors are included in a separate section of this chapter. However, other parts of the chapter are independent of this portion.

To conserve space and retain simplicity, the cross-field theory of single-phase motors has been omitted. But trade and technical people will find sufficient information for their needs. Additional material on speed control of single-phase motors can be found in Chapter 9.

SINGLE-PHASE SQUIRREL-CAGE MOTORS

Construction

The rotor of a single-phase squirrel-cage motor is essentially the same as that used in three-phase motors and needs no further description. Except for shaded-pole types, the stator core is also very similar, but a single-phase stator may have slots of two or three different sizes, whereas in three-phase machines, the slots are normally all the same size. A single-phase stator can be wound for any even number of poles, two, four and six being most common. Like three-phase machines, adjacent poles have opposite magnetic polarity, and Equation 4–1 regarding synchronous speeds applies. The stator windings, however, are distinctively different in two ways.

First, single-phase motors usually have concentric coils, as shown in Figure 7-1. In that diagram the coils appear to have only one turn of wire, but in practice each coil will have many turns. With concentric coils, the designer can adjust the number of turns of wire in each coil to obtain (as nearly as possible) sinusoidal distribution of mmf along the air gap, and that is one of their advantages.

Second, single-phase squirrel-cage motors normally have two stator windings. Except for shaded-pole motors, the two windings are similar to Figure 7-1, but one of them usually

FIGURE 7-1   Four-pole single-phase motor winding

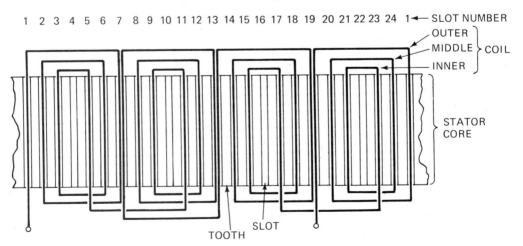

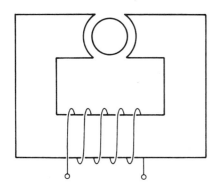

FIGURE 7-2 Simplified magnetic circuit for a two-pole motor

has fewer turns of much smaller wire than the other, and the two windings are shifted 90° electrical from each other. In motors that operate with both windings energized, the winding with the heaviest wire is known as the main winding, and the other is called the auxiliary winding. If the motor runs with the auxiliary winding open, we generally call them the running and starting windings, respectively. In most motors, the main winding is installed first and tamped down into the bottom of the slots. The starting winding is then placed in the slots so that it lies on top of (but shifted 90° from) the running winding.

If it has four or more poles, the main winding of a shaded-pole motor is similar to the windings that have already been described, except that it generally has only one coil per pole. If it has only two poles, a shaded-pole motor may be constructed similarly to the magnetic circuit shown in Figure 7-2.

Counterrotating Field Theory

If we consider the main winding only in a two-pole motor or examine the magnetic circuit in Figure 7-2, it is apparent that the main winding can develop an mmf along only one axis (horizontal in Figure 7-2). We can regard the mmf as being stationary in space and varying sinusoidally with respect to time. However, the sum of two equal-strength phasors rotating at equal speeds in opposite directions would also alternate along one axis as shown in Figure 7-3. We therefore can regard the mmf of a single stator winding as having two equal-strength components rotating in opposite directions — and that is the key to single-phase motor operation.

If the air gap is uniform and the rotor is at a standstill, the counterrotating components of the stator mmf create equal-strength counterrotating magnetic field components. Both of these components generate a cemf in the stator winding, and both components affect the squirrel cage equally so that the net torque is zero. The mmf of the squirrel-cage currents at standstill also opposes both field mmf components equally.

If the rotor is given a spin in a direction that we will call forward, three things happen. First, if X_{BA} is several multiples of R_A so that the squirrel cage develops maximum torque at moderate values of slip, the torque produced by the forward-rotating field becomes greater than that produced by the backward field, and so a net forward torque is produced.

Second, the induced currents in the rotor tend to suppress the backward field more than the forward field, hence increasing the net torque. If the rotor is turning at synchronous speed, the backward field is almost eliminated.

Third, since cemf is produced by both field components and must remain fairly constant, suppression of the backward field results in a stronger forward field, the sum of the two remaining roughly constant. This also increases the net torque.

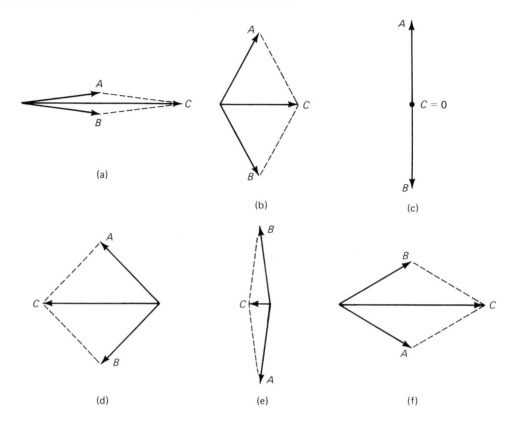

FIGURE 7-3 Two counter-rotating phasors and their sum

The torque developed by the forward and backward fields and the net torque of a single-phase motor are shown graphically in Figure 7-4. it is easy to see why a motor with only one stator winding will run in whichever direction it is initially started, but let us consider three other implications of the counter-rotating field theory.

First, if a motor runs with only one stator winding, it is important that the rotor resistance be much less than its blocked rotor reactance. Increasing the rotor resistance increases the effectiveness of the backward field, which reduces breakdown torque, lowers efficiency, and causes increased slip. Double squirrel cages and deep narrow rotor bars are also impractical for the same reasons, and so high starting torque must be obtained some other way.

Second, when the motor is running, the presence of the backward field creates torque pulsations on the rotor. This is why single-phase motors tend to be somewhat noisier than three-phase types. Third, if we could suppress the backward field, motor performance would be improved.

MMF of Two Stator Windings

If we have two identical windings 90° electrical apart and carrying equal currents 90° out of phase with each other, we can show with phasors that the two windings together produce a constant-strength, smoothly rotating mmf. However, if we split the mmf of each winding into forward and backward components, we can also show that their forward components add together, while their backward components cancel each other out. If K_{RW} and K_{SW} represent the peak mmf's of the two windings, the net forward component becomes $(K_{RW} + K_{SW})/2$, and the backward components become $(K_{RW} - K_{SW})/2$. Even if the currents are not 90° out of phase (as long as they are not in phase with each other), or for that matter even if the windings are not 90° apart (as long as they are not superimposed), the phasor sum of the forward mmf's always exceeds the sum of the backward mmf's, and so a net rotating mmf is produced.

The reason for having two stator windings in a single-phase motor is to create a rotating magnetic field in the motor. We need the rotating field to start the motor, and in some motors both windings are energized under normal running conditions.

As used here, the forward and backward

directions are purely arbitrary. It is important to note that if the mmf of one winding is reversed, the net field will rotate in the other direction. This is how most squirrel-cage single-phase motors are reversed.

Types of Single-Phase Squirrel-Cage Motors

Split-Phase Motor

The starting winding in these motors has wire of about one quarter the cross section of the wire in the running winding. This gives the starting winding a smaller X/R ratio, and so the current in the starting winding (I_{SW}) leads the current in the running winding (I_{RW}), and a rotating field is produced. The angle between the two currents is usually about 30°, as shown in Figure 7-5, and so the starting torque is only fair. These motors normally have a centrifugally operated starting switch that cuts out the starting winding when the motor reaches about 75% of synchronous speed and recloses at about 25%. The torque curve therefore appears as a loop, as shown in Figure 7-5.

The starting winding in these motors heats rapidly during the starting period. If the load has too much inertia, or if the starting switch fails to open, the starting winding will quickly burn out.

Capacitor Start Motor

The wire used for the starting winding has about half the cross section of the wire used for the running winding, and an electrolytic capacitor is wired in series with the starting winding. The capacitor results in a greater phase angle between the currents in the two windings, and so the motor develops good starting torque using less starting current than a split-phase motor. A schematic, a phasor diagram, and a torque speed curve

FIGURE 7-4 Torque produced on a squirrel cage by a single stator winding

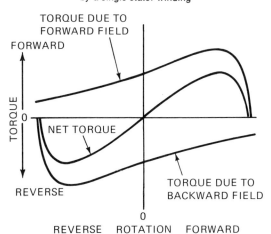

TORQUE DUE TO FORWARD FIELD

FORWARD

TORQUE

0

NET TORQUE

REVERSE

TORQUE DUE TO BACKWARD FIELD

0

REVERSE   ROTATION   FORWARD

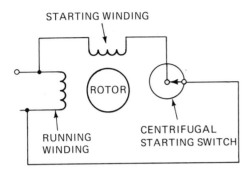

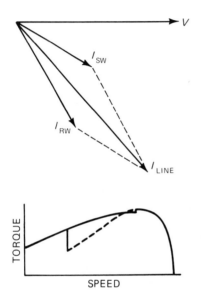

FIGURE 7-5   Split-phase motor

start motors, but in addition to the electrolytic capacitor, they have a continuous-duty oil-filled capacitor of somewhat smaller capacitance (microfarads). The starting switch cuts out the electrolytic capacitor when the motor reaches about 75% of synchronous speed, and the motor runs with both windings energized, as shown in Figure 7-7. The torque curve is much like that of a capacitor start motor, but the breakdown torque is usually higher. Under locked rotor conditions the phasor diagram is also like that

FIGURE 7-6   Capacitor-start motor

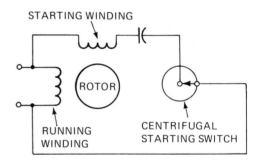

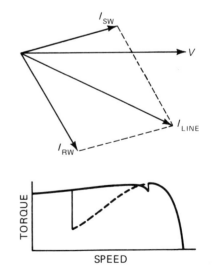

for this motor are shown in Figure 7-6. The electrolytic capacitor is generally for short time duty only, and starting winding burn-out can still occur; but for high-inertia or high starting-torque loads, the capacitor start motor is much better than a split-phase type.

Two-Value Capacitor Motor

Also known as capacitor start–capacitor run motors, these are similar to capacitor

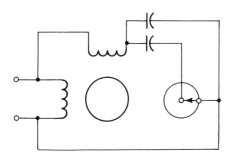

FIGURE 7-7   Two-value capacitor motor

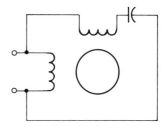

FIGURE 7-8   Permanent-split capacitor motor

shown in Figure 7-6, but under running conditions, the starting winding current is somewhat reduced.

In addition to raising breakdown torque, keeping both windings energized improves motor performance in two ways. First, the backward field components are reduced so that the motor runs with less vibration. Second, the leading current in the starting winding improves the overall power factor and reduces the full-load line current.

Permanent Split Capacitor Motor

The auxiliary winding in these motors usually has almost the same size of wire and almost as many turns as the main winding; in fact, on some motors the two windings are identical. A continuous-duty oil-filled capacitor is connected in series with the auxiliary winding, and as shown in Figure 7-8, there is no starting switch.

The capacitor size is chosen so that under running conditions the currents in the two windings are about 90° apart, but the actual angle varies somewhat with changes of load. Under starting conditions the capacitor is too small, and so the starting torque tends to be low; but the manufacturer can partly compensate for this by using a higher-resistance squirrel cage.

A permanent split capacitor motor runs very quietly and operates with high power factor and low line current. Because it has no starting switch, it is suitable for wide-range speed control.

Shaded-Pole Motor

As shown in Figure 7-9, a shaded-pole motor has a bare copper loop enclosing an axial strip about one quarter of a pole width along one side of each pole. Flux from the main winding induces currents (I_{SR}) in the shading coils (or shading rings as they are sometimes called), and these currents always lag the current in the main winding, as shown in Figure 7-9. As a result, the flux moves from the unshaded to the shaded portions of each pole in each alternation; that is, a rotating magnetic field is produced.

Most shaded-pole motors take advantage of another starting method known as reluctance starting. Note in Figure 7-9 that the air gap is shown as being wider on the pole sides opposite the shading rings. The flux in the narrow air gap tends to lag the flux in the wide air gap, and so a rotating field is produced, moving from the wide to the narrow air gap. This improves starting torque.

Economy of manufacture is the chief virtue of a shaded-pole motor. It can be used for wide-range speed control, but its efficiency and power factor are very poor.

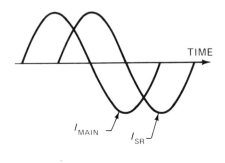

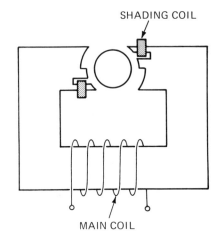

FIGURE 7-9 Shaded-pole motors

OTHER TYPES OF SINGLE-PHASE MOTORS

Universal Motor

A universal motor is quite similar to a series-wound dc motor but is specifically designed to run on direct or alternating current of the same voltage and any frequency up to and including 60 Hz. The main design changes are the use of a minimal number of turns in the field coils and the practice of laminating the entire magnetic circuit.

A typical universal motor is shown in Figure 7-10. Some universal motors have the main field winding distributed in slots (similar to split-phase motors), and some have compensating winding (also distributed) set at right angles to the main field.

FIGURE 7-10 Universal motor

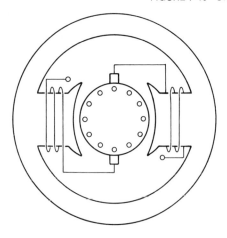

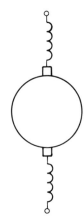

Looking at Figure 7-10, it is apparent that if the instantaneous line voltage reverses, the field flux and armature conductor currents will both reverse, and therefore the torque remains in the same direction. Reversed rotation can be obtained only by reversing the armature connections relative to the series field coils.

When operated on alternating current, saturation of the magnetic circuit during the peak current times of the cycle tends to raise the motor speed. However, inductive reactance tends to choke out the motor current (and therefore reduces motor torque) considerably on alternating current, and so the motor performs somewhat better on direct current. Speed control can be obtained either by changing the applied voltage or by adding resistance in series.

The torque curve of a universal motor is similar to a series-wound dc motor with high starting torque and a high no-load speed. With no load, it may run away, and so direct coupling or gear drives are usually employed.

The universal motor is usually designed to run at very high speeds, often 200 to 400 rev/s. Because they can run at such high speeds, these motors are smaller and lighter than other motors of the same output, and this makes them particularly suitable for hand-held portable tools.

Repulsion Family of Motors

The stator winding of these motors is practically the same as the main winding of a split-phase motor. The rotor has a winding essentially the same as a lap-wound dc armature. However, there is no electrical connection between the armature and the line. Instead, the brushes in these motors are short circuited together, and the flux produced by the stator winding induces current in the armature. There are three types of motors in this group.

Repulsion Motor

A cross-sectional view of a repulsion motor appears in Figure 7-11. The crosses and dots in the armature conductors show the relative direction of the induced voltages in the armature windings. If the brushes are positioned as shown, the voltage between the brushes becomes zero (the voltages in half of the conductors in each path oppose the voltages in the other half of that path), and so there will be no armature current. With voltage applied to the stator, the motor hums softly and produces no torque. We therefore call this brush position the "soft-neutral" position. If we move the brushes away from soft neutral, we get armature current and torque will be produced. Maximum torque occurs when the brushes are about 65° to 75° away from soft neutral. If the brushes

FIGURE 7-11 Repulsion motor

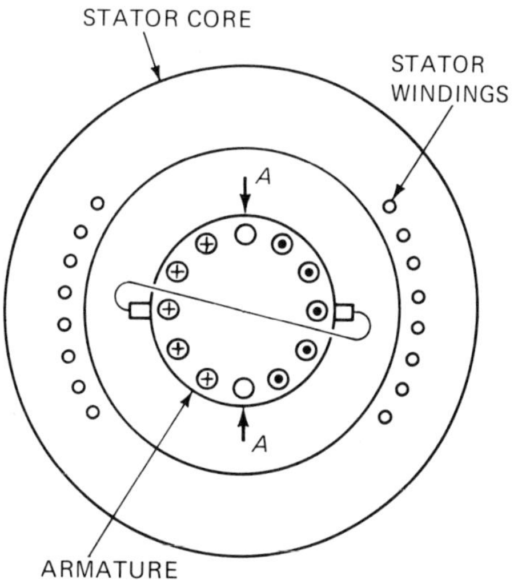

are moved 90° from soft neutral (points *A* and *A'*), armature current will be high, but the torque is again zero and the motor growls harshly. We call this brush position the "hard-neutral" position.

With any given brush position, the torque–speed curve looks much like that of a universal motor. However, we can adjust the height of the curve by moving the brushes and in that way obtain speed control. These motors can run at high speeds; 125 rev/s is not unusual. Rotation can be reversed by moving the brushes the opposite way from the soft-neutral position.

The repulsion motor has never enjoyed much popularity. Its chief use appears to have been on applications requiring speed control.

Repulsion Start Induction Motor

This is basically a repulsion motor, but it has a centrifugally operated mechanism that short circuits all the commutator bars together when the motor reaches about 75% of synchronous speed. The armature coils then act like a squirrel cage, and so the motor has good speed regulation and a very definite no-load speed.

To reduce noise and prolong brush life, some of these motors have a brush-lifting mechanism that raises the brushes off the commutator when the motor comes up to speed. The direction of rotation is reversed by moving the brushes.

The repulsion start induction (RSI) motor has tremendous starting torque (600% is not unusual), and it will accelerate high-inertia loads without distress. Unfortunately, it is a very expensive motor to produce, and so has been almost completely superseded by the capacitor start motor.

Repulsion Induction Motor

These are similar to a repulsion motor, but underneath the repulsion winding the armature has a squirrel cage. The repulsion winding provides good starting torque, with the squirrel cage also contributing torque as the motor accelerates. However, at no load the repulsion winding tries to run away and is held back only by the squirrel cage. As a result, the no-load speed is above synchronous and the no-load current rather high, sometimes higher than the full-load value. Some of these motors will run away if the brush position has been wrongly adjusted. The repulsion induction (RI) motor is also a very expensive motor to manufacture and few have been produced in Canada or the United States since about 1940.

Synchronous Motors

A synchronous motor is one that runs at synchronous speed, that is, with no slip. For many applications the slip of an induction motor is of little consequence; but for devices like electric clocks, tape drives, and record player turntables, slip would permit variations of speed that cannot be tolerated.

Most single-phase synchronous motors use shaded-pole stators and rotors that can lock into step with the rotating magnetic field. Usually, they are very small, of less than 15-W output. There are two common types.

Hysteresis Motor

The rotor is made of hard steel that has appreciable hysteresis. If it is not turning at synchronous speed, the rotor is continually being remagnetized in a new direction, and because of hysteresis, torque is developed on the rotor. When it reaches synchronous speed, the rotor becomes essentially a permanent magnet, and its poles lock into step with the magnetic poles of the stator.

Reluctance Motor

The rotors are similar to squirrel-cage rotors but have portions cut away, leaving projecting iron poles as shown in Figure 7-12. The motor starts as a squirrel-cage machine, but when it gets very close to synchronous speed the rotor poles are attracted to the stator poles and lock into step with them. The motor then runs at synchronous speed.

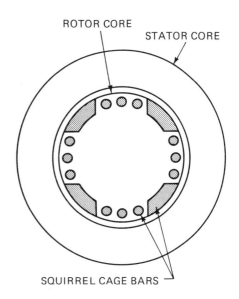

FIGURE 7-12 (right)   Reluctance motor

SOME ASPECTS OF STATOR WINDING DESIGN

Reversible Dual-Voltage Motors

To minimize the number of different designs that must be produced, manufacturers tend to build their motors so that the rotation can be easily reversed and the motor can be operated at either 115 or 230 V. The following are some of the more common types. Unless otherwise indicated, the numbers given for the leads and the connecting instructions are in accordance with NEMA and EEMAC standards.

Figure 7-13 shows a reversible dual-voltage capacitor start motor. The starting winding and both halves of the running winding are rated at 115 V, so on the high-voltage connection, the starting winding is connected in parallel with one half of the running winding. Split-phase and two-value capacitor motors follow the same diagram.

Figure 7-14 shows a reversible dual-voltage

FIGURE 7-13   Reversible dual-voltage capacitor-start motor

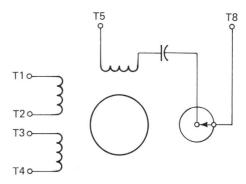

VOLTAGE	CONNECT	
LOW	T1, T3 + T5 → L$_1$	TO REVERSE INTERCHANGE T5 AND T8
	T2, T4 + T8 → L$_2$	
HIGH	T1 → L$_1$	
	T2, T3 + T5	
	T4 + T8 → L$_2$	

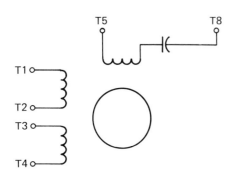

VOLTAGE	CONNECT	
LOW	T1, T3 + T5 → L₁	
	T2, T4 + T8 → L₂	TO REVERSE
HIGH	T1 + T5 → L₁	INTERCHANGE
	T2 + T3	T5 AND T8
	T4 + T8 → L₂	

FIGURE 7-14   Reversible dual-voltage permanent-
split capacitor motor

permanent split capacitor motor.[1] As indicated by the connecting instructions, the
auxiliary winding in at least some of these
motors can be connected directly across a
230-V line.

Reversible Single-Voltage Motors

For single-voltage split phase, capacitor
start, two value capacitor, and some permanent split capacitor motors, the schematic
diagrams are similar to those in Figures
7-13 and 7-14, but leads 2 and 3 do not exist
(think of them as being joined inside the
motor). Only the external leads 1, 4, 5, and
8 remain.

Some permanent split capacitor motors
have identical main and auxiliary windings

---

[1]This motor does not appear to be in the NEMA
standards.

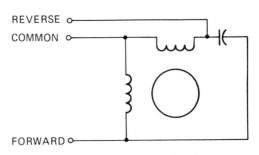

FIGURE 7-15   Reversible single-voltage
permanent-split capacitor motor

and are wired as shown in Figure 7-15. In
these motors, reversed rotation is obtained
by interchanging the roles of the main and
auxiliary windings.

In most cases, reversed rotation of a
shaded-pole motor can be obtained by disassembling the motor and then reassembling
it with the stator core reversed relative to
the shaft extension, but this is practical only
for unidirectional drives. If it must be able
to run in either of two directions, the motor
can be built with two sets of shading coils,
one coil on each side of each pole. The
shading coils in this case must be wound
with insulated wire and provided with external leads, as shown in Figure 7-16. Close
the appropriate set of shading coils to get
the desired direction of rotation.

For servomechanisms, shaded-pole motors

FIGURE 7-16   Reversible shaded-pole motor

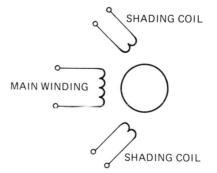

similar to that in Figure 7-16 are some-times used. However, these are designed so that the main winding can be left energized indefinitely under locked rotor conditions, and the rotor has enough resistance that it will not run unless one set of shading rings is closed. This lends itself to simple control; a single-pole double-throw center-off control switch (or its equivalent) is all that will be required.

Integral Thermal Protection

Many motors have built-in protection against overheating. The most common arrangement consists of a small resistor or heating element and a pair of normally closed contacts ac-tuated by a bimetallic disc. The resistor is positioned so that it heats the disc, and since the whole assembly is inside the motor, the temperature rise of the motor also tends to open the contacts. The resistor is connected in series with the motor, so this device is sensitive to both motor temperature and current.

A reversible single-voltage motor with integral thermal protection is shown in Figure 7-17. One type of reversible dual-voltage motor with thermal protection is shown in Figure 7-18, but other arrange-ments are also recognized by NEMA.

FIGURE 7-17   Reversible single-voltage motor
with thermal protection

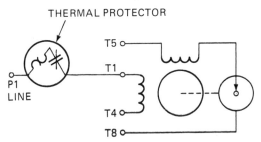

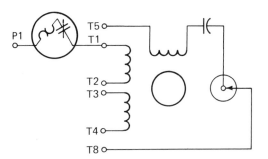

FIGURE 7-18   Reversible dual-voltage motor with
thermal protection

Two-Speed Motors

A two-speed shaded-pole motor for use with fan loads is shown in Figure 7-19. Use of all the existing turns in the winding reduces the voltage per turn, which decreases the motor torque. The motor will then run at reduced speed, but the actual speed obtained depends on the magnitude of the load. At no load there will be no appreciable change of speed. This scheme can be used to pro-vide more than two speeds.

Two-speed permanent split capacitor mo-tors are rare, but starting switch types of two-speed single-phase motors are fairly common. Because the starting switch is cen-trifugally operated, the ratio between the high and low speeds seldom exceeds 2 to 1. If the motor speeds are in a 2 to 1 ratio, the main winding will be a pole-changing design

FIGURE 7-19   Two-speed shaded-pole motor

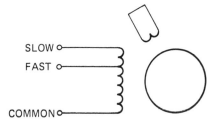

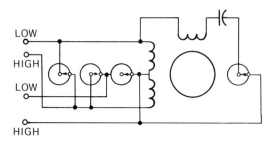

FIGURE 7-20 Two-speed consequent-pole motor

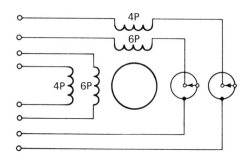

FIGURE 7-21 Two-speed motor with two starting
windings

that operates as a conventional winding with a parallel (center tap to both ends) connection and a consequent-pole winding with the series connection (e.g., a four-to-eight pole arrangement). The starting winding will not be a pole-changing type, but instead is designed only for the lower number of poles. A schematic diagram of such a motor is shown in Figure 7-20. The motor starts on its high speed, and (if the low-speed terminals have been energized) the starting switch disconnects the starting winding and changes the main winding to the consequent-pole arrangement when the motor comes up to speed.

If the two speeds are not in a 2 to 1 ratio, the motor will have two main windings, usually four and six pole, respectively. Some of these motors have two starting windings,

one for each speed, as shown in Figure 7-21. Others have only one starting winding that works in conjunction with the high-speed winding, as shown in Figure 7-22. This latter type of motor always starts on high speed. When the motor comes up to speed, the starting switch disconnects the starting winding, and if the low-speed terminals have been energized, it disconnects the high-speed main winding and connects the low-speed main winding to the line. The terminal markings of two-speed motors apparently have not yet been standardized by NEMA.

Starting Relays

The centrifugally operated starting switch has three disadvantages. First, it is probably the

FIGURE 7-22 Two-speed motor with one starting winding

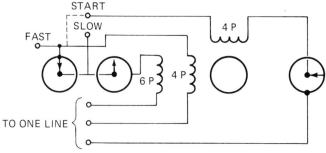

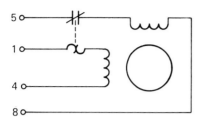

FIGURE 7-23 Thermal starting relay

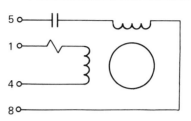

FIGURE 7-24 Current-activated starting relay

most troublesome part of a single-phase motor. Second, it prevents plug-stop operation[2] of the motor. Third, for hermetically sealed refrigeration compressors (where the motor is enclosed in a refrigerant such as Freon), the arc that would occur at the switch contacts would contaminate the refrigerant. So for motors that will be very inaccessible (such as a pump motor at the bottom of a well), for applications that require plugging (such as an electrically driven chain hoist), and for hermetically sealed compressor motors, external starting relays are used instead.

A thermal starting relay is shown in Figure 7-23. When the motor is energized, the heater element opens the contacts after about 2 seconds, cutting off the starting winding. A thermistor connected in series with the starting winding will do the same thing. These devices are popular with the air-conditioning industry. However, they must be allowed to cool before they will reclose the starting winding circuit and so are not suitable for plugging or quick start–stop–restart kinds of operation.

A current-actuated starting relay is shown in Figure 7-24. When the motor is initially

energized, the inrush current picks up the starting relay and cuts in the starting winding. When the motor reaches full speed, the current drops off and the relay releases, deenergizing the starting winding. These relays are suitable for plugging and repetitive starting and can be designed for any switch-start motor.

A voltage-actuated starting relay is shown in Figure 7-25. On capacitor start motors the voltage across the starting winding rises sharply as the motor approaches synchronous speed, and this opens the starting relay. When the motor is running, the cemf generated in the starting winding keeps the relay open.

These relays are suitable for plugging and repetitive starting, but cannot be applied to a split-phase motor.

FIGURE 7-25 Voltage-actuated starting relay

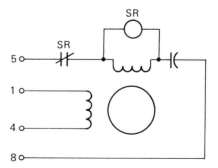

[2]Plugging or plug-stop operation is the practice of stopping a motor by temporarily connecting it to run with reversed rotation.

MATHEMATICAL INTRODUCTION TO SINGLE-PHASE MOTORS

A mathematical treatment of all types of single-phase motors is too extensive to include here, so we will confine ourselves to those squirrel-cage types that have two stator windings displaced 90° from each other (i.e., split-phase, capacitor start, permanent split, and two-value capacitor motors). We will consider blocked-rotor conditions with both windings energized and running conditions with one winding energized and the other one open, but we will not examine running conditions with both windings in use.

Starting Torque of Single-Phase Motors

The starting torque of the single-phase motor types we are considering is given by Equation 7–1.

$$T = K_T I_{RW} I_{SW} \sin \angle_{I_{RW}}^{I_{SW}} \qquad (7\text{--}1)$$

where K_T is a constant that can be determined from the physical dimensions of the motor.

Let us now consider the design problem of obtaining maximum starting torque from a split-phase motor. Specifically, we will assume that all other factors will remain unchanged, and the designer wishes to adjust the starting winding resistance (by changing the wire size) to get maximum starting torque. Increasing the resistance of the starting winding will decrease the starting winding current and bring it more nearly in phase with the voltage; in fact, the locus of I_{SW} is a semicircle, as shown in Figure 7-26. The quantity $\left(I_{SW} \sin \angle_{I_{RW}}^{I_{SW}} \right)$ from Equation

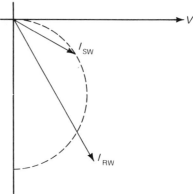

FIGURE 7-26 Maximizing the starting torque of a split-phase motor

7–1 is just the perpendicular distance from the phasor I_{RW} to the tip of I_{SW}, and it can be shown that this is at a maximum when

$$\angle_{I_{SW}}^{V} = \tfrac{1}{2} \angle_{I_{RW}}^{V}$$

If the starting winding circuit includes a capacitor, the designer can adjust the starting torque by changing the capacitor size. The locus of the starting winding current will then be as shown in Figure 7-27. The phasor *A* is the starting winding current required to obtain maximum torque, and phasor *B* shows the value of I_{SW} that will give maximum torque per line ampere.

Pitch and Distribution Factors

The use of concentric coils in a motor makes the distribution factor equal to unity, and so it does not appear in our equations. However, since the coils do not all have the same span, we have to determine separate pitch factors for each size of coil using Equation 4–9 and if necessary Equation 2–4.

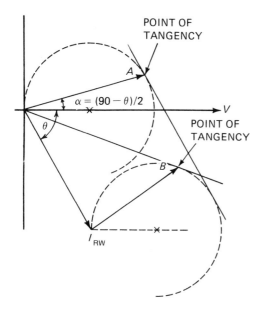

FIGURE 7-27   Maximizing the starting torque of a
capacitor-start motor

Let us identify the coils in one pole as 1, 2, 3, etc., so that n_1, n_2, n_3, etc., will represent the number of turns of wire in each coil and K_{p1}, K_{p2}, K_{p3}, etc., will represent the pitch factors for those coils. The effective number of conductors (n_s) in the stator can then be determined using Equation 7-2, where p equals the number of poles.

$$n_s = 2p(K_{p1}n_1 + K_{p2}n_2 + K_{p3}n_3 \\ + \cdots + K_{pn}n_n) \quad (7\text{-}2)$$

Flux and Counter EMF

The relationship between counter emf and the maximum flux per pole (ϕ_p) is given by Equation 7-3, which is valid under locked rotor conditions or when running with only one stator winding energized.

$$E_s = 2.22fn_sP\phi_p \quad (7\text{-}3)$$

The product $P\phi_p$ gives the total air gap flux

from which we can calculate B_{max} if desired, using Equation 4-20.

Equivalent Circuits for Single-Phase Motors

Under blocked-rotor conditions, the equivalent circuit for each stator winding of a single-phase motor is shown in Figure 7-28. If the resistance and reactance per bar (R_A and X_{BA} of the squirrel cage) are known, R_R and X_R can be determined from Equations 7-4 and 7-5. Blocked-rotor iron losses and magnetizing current are accounted for by G_o and B_o, respectively.

$$R_R = \frac{2n_s^2}{N_A}R_A \quad (7\text{-}4)$$

$$X_R = \frac{2n_s^2}{N_A}X_{BA} \quad (7\text{-}5)$$

If only one stator winding is energized, it can be represented by the circuit in Figure 7-29, under both locked rotor and running conditions. The currents I_{RF} and I_{RB} represent the squirrel-cage currents due to the forward and backward fields, respectively. The total power in the conductances ($2G$) is still the total iron losses. Rotor copper loss (RCL) and rotor power developed (RPD) are given

FIGURE 7-28   Equivalent circuit for one winding
of a single-phase motor (blocked rotor)

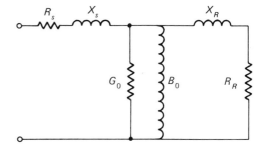

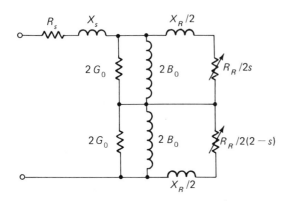

by Equations 7–6 and 7–7, respectively. Other calculations are essentially the same as those that apply to three-phase motors.

$$RCL = 0.5R_R(I_{RF}^2 + I_{RB}^2) \qquad (7\text{–}6)$$

$$RPD = I_{RF}^2\left(\frac{0.5R_R}{s}\right) - I_{RB}^2\left(\frac{0.5R_R}{2 - s}\right) \quad (7\text{–}7)$$

FIGURE 7-29 (left)   Equivalent circuit for a single-phase motor (only one stator winding energized)

SUMMARY

The majority of single-phase motors are of relatively small size. However, they have many applications, mostly in domestic appliances and portable tools. Single- and three-phase squirrel-cage motors have many points of similarity. The rotors and other mechanical parts are physically almost indistinguishable, and only the stator windings are distinctive. The equivalent circuit for the main winding of a split-phase or capacitor start motor is like one phase of two three-phase motors that are mechanically coupled together and electrically connected in series so as to run in opposite directions. However, motors that run with two stator windings energized are not so easily described.

The universal motor is similar to the series-wound dc motor, and its use is almost completely confined to portable tools. Like the synchronous types, most universal motors are sold to original-equipment manufacturers (OEM users) who build them into their final product (such as clocks, tape recorders, and electric drills). Other types of single-phase motors exist but are becoming increasingly rare because of high manufacturing costs.

REVIEW QUESTIONS

7-1.   Explain what is meant by the following terms:
   (a) concentric coils
   (b) forward rotating field component
   (c) split-phase motor
   (d) capacitor start motor
   (e) shading coil
   (f) reluctance starting
   (g) universal motor
   (h) synchronous motor
   (i) hysteresis motor
   (j) reluctance motor
   (k) thermal protection

7-2.   Motors used in residential areas are usually single phase. Why?

7-3.   What is the most common visual difference between single- and three-phase stator cores that have no winding in them?

7-4.   What is the advantage of using concentric coils rather than lap coils in single-phase motors?

7-5. Briefly write out in your own words an explanation of "the counterrotating field theory of single-phase motor operation."

7-6. How do squirrel-cage types of single-phase motors start?

7-7. To the uninitiated, split phase, capacitor start, two-value capacitor and permanent split capacitor motors all look alike. How can each one be visually distinguished from the others?

7-8. When started and stopped, many single-phase motors make a distinct click sound as they approach full speed and also as they coast to a stop. Which motors make this click noise and what internal parts produce it?

7-9. Why is a split-phase motor not suitable for repetitive starting of high-inertia loads?

7-10. Manufacturers of window air conditioners find it desirable to use the most powerful compressor motor that can be operated from a 15-A branch circuit. What types of single-phase motors would be desirable?

7-11. Explain how shaded-pole starting works.

7-12. How can the universal motor be visually distinguished from the repulsion family of motors?

7-13. A universal motor usually performs better on direct than on alternating current. Why?

7-14. Why is the universal motor so popular for hand-held portable tools?

7-15. Carefully distinguish between the three members of the repulsion family of motors.

7-16. Distinguish between the two types of single-phase synchronous motors.

7-17. Capacitor start motors sold to original-equipment manufacturers (OEM users) for inclusion in their products tend to be single-voltage designs that are not easily reversed, but those sold directly to the ultimate user are frequently dual-voltage reversible types. Why?

7-18. Write out telephone instructions (i.e., no diagrams) explaining the following:
   (a) How to hook up a dual-voltage capacitor start motor for each voltage (assume no thermal protection).
   (b) How to reverse the rotation of the same motor.

7-19. Do the correct answers for question 7-18 also apply to the following dual-voltage motors; if not, what are the differences?
   (a) Split-phase motor (no thermal protection).
   (b) Two-value capacitor motor (no thermal protection).
   (c) Permanent split capacitor motor (no thermal protection).
   (d) Capacitor start motor with thermal protection.

7-20. Even if the rotation of a shaded-pole motor can be reversed, it might not be suitable for a load whose direction of rotation must be frequently reversed. Why?

7-21. Explain how integral thermal protection devices work.

7-22. For some applications of two-speed motors, the operator must be able to select either direction of rotation at either speed. Draw schematic diagrams to show how this can be done for each of the three types of two-speed motors. Use two double-throw switches for each, one marked "fast–slow," one marked "forward–reverse."

7-23. What are the three disadvantages of centrifugally operated starting switches?

7-24. Explain how a thermal starting relay works and why it is not suitable for plugging applications.

7-25. Current-actuated and voltage-actuated starting relays can be distinguished by noting the contact position when the relay is deenergized. Explain.

7-26. Explain how a current-actuated starting relay operates.

7-27. Explain how a voltage-actuated starting relay operates.

7-28. Why do capacitor start motors develop more starting torque with less starting current than split-phase motors?

7-29. Why do two-value capacitor motors have higher starting torque than permanent split capacitor types?

7-30. Why do split-phase motors generally draw more line current than a permanent split capacitor motor of the same output?

MATHEMATICS PROBLEMS

7-1. Find the synchronous speeds of two-, four-, six-, and eight-pole motors at both 60 and 50 Hz.

7-2. Find the full-load per unit slip of each of the following motors. The speeds quoted are full-load values.
   (a) 60 Hz, 57.5 rev/s
   (b) 50 Hz, 47.5 rev/s
   (c) 4 pole, 60 Hz, 28.75 rev/s
   (d) 6 pole, 60 Hz, 17.5 rev/s
   (e) 8 pole, 50 Hz, 11.25 rev/s

7-3. The running winding of a single-phase motor has been designed so that under blocked rotor conditions with 115 V applied its current will be 20 A at a power factor of 0.4225 lagging. Under the same conditions the starting winding current is 18 A lagging the voltage by 40°, and the starting torque is 1.4 newton-meters.
   (a) Find the starting (line) current of the motor.
   (b) To get maximum starting torque, how much resistance should be added to the starting winding, and what values of starting (line) cur-

rent and starting torque will then be obtained?
   (c) Instead of adding resistance, the designer chooses to add a capacitor in series with the original starting winding. Assuming 60 Hz, (i) what size (microfarads) capacitor gives maximum starting torque, and what values of starting torque and line current result? (ii) What size (microfarad) capacitor gives maximum starting torque per line ampere, and what values of starting current and starting torque result?

7-4. A 36-slot stator core that is 6 cm long and has a bore diameter of 7.5 cm is used to build four- and also six-pole, 115-V, 60-Hz motors. If the winding specifications are as shown in the table, find the following:
   (a) The effective number of stator conductors in each winding.
   (b) The flux per pole and also the maximum flux density if only the main winding is energized (assume sinusoidal flux distribution).

Coil	Main Winding		Starting Winding	
	Span (slots)	Turns	Span (slots)	Turns
FOUR POLE				
Inner	5	21	4	24
Middle	7	50	6	48
Outer	9	28	8	24
SIX POLE				
Inner	2	20	2	28
Middle	4	58	4	56
Outer	6	30	6	28

7-5. The rotors used for the motors in problem 7-4 have 31 bars and each bar has 52.5-$\mu\Omega$ resistance and 0.388-μH inductance. Find the reflected values of resistance and reactance (X_R and R_R) for both the starting and running windings in each motor.

7-6. For simplicity, let us neglect the iron losses and magnetizing current of the motors in problems 7-4 and 7-5 (i.e., G_o and B_o become zero). If the main stator winding resistance and reactance (R_s and X_s) are 1.5 and 1.0 Ω, respectively, for the four-pole design, and 2.3 and 1.4 Ω, respectively, for the six-pole motor, calculate the torque developed by each motor at 0, 5, 10, 15, 25, 40, 70, and 100% slip, and graph the results (assume that the starting winding is open).

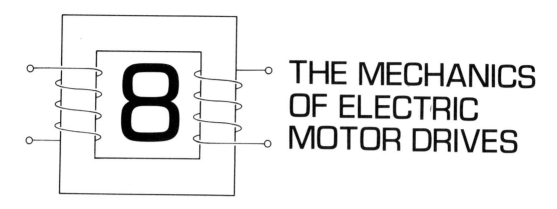

THE MECHANICS OF ELECTRIC MOTOR DRIVES

Sir Isaac Newton's law of motion states that the acceleration of a body is directly proportional to the net force applied, and inversely proportional to the mass. The selection of a motor with suitable characteristics and to some extent the choice of a starter or controller for a given installation both hinge on this basic principle, and so do the speed-control capabilities and speed-regulation characteristics of all the various motor and load combinations. For economic reasons it is desirable to avoid the use of a motor that has an unnecessarily large power rating. So we will therefore examine electric motor drives from a mechanical viewpoint and show how the mechanical characteristics of the load determine the choice of motor power and torque characteristics and the selection of appropriate control equipment.

To proceed logically, we will first discuss the torque–speed and power–speed characteristics of typical loads. The selection of a motor of the correct power rating will be our second topic. Then we will consider Newton's law of motion as applied to motor drives, including acceleration of the load, the effects of load inertia, and the mechanics of speed regulation and speed control.

There are other important factors that must be considered for any motor application, such as the choice of motor speed, selection of the drive method (direct coupling, belt, chain, or gear drives), mounting position, type of enclosure, and the like, but we will not go into these topics here.

TORQUE AND POWER REQUIREMENTS OF MACHINES

Since power is proportional to the product of torque and speed, it is convenient to examine simultaneously the torque and power requirements of various loads. The

199

power and torque required to drive a given machine at a constant speed are usually required for two reasons:

To Overcome Friction

Although it is often only a small part of the total load, friction accounts for the entire load of some machines (e.g., a horizontal conveyor belt), and it appears in three types:

Coulomb (Mechanical or Sliding) Friction

In a given machine coulomb friction tends to be constant regardless of speed, except that it is usually higher under static conditions than it is when the machine is running. The torque and power required to overcome coulomb friction are shown in Figure 8-1.

Viscous (Fluid) Friction

The torque required to overcome viscous friction is proportional to the square root of the speed so that the power required is proportional to $(rev/s)^{3/2}$, as shown in Figure 8-2. This is the type of friction produced

by parts like sleeve bearings under running conditions.[1]

Rolling Friction

Many factors affect the coefficient of friction in rolling-type bearings, so a simple torque–speed relationship does not exist.

To Perform Useful Work

Mere rotation of the driven machine is seldom the main objective of a motor drive. The real purpose is to have the machine do useful work on some processed material or series of objects, which we will call the throughput. The higher the throughput rate, the more torque (and power) will be required to drive the machine.

On many machines there is an inherent relationship between operating speed and throughput, and so (with other factors remaining constant) there is an inherent relationship between operating speed and the driving power and torque required. The exact nature of this relationship depends upon the

[1] Under static conditions, a sleeve bearing loses its oil film and reverts to coulomb friction, which is much higher.

FIGURE 8-1   Torque and power required for coulomb friction

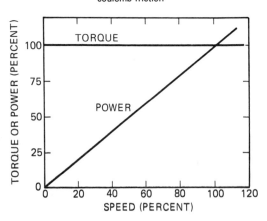

FIGURE 8-2   Torque and power required for viscous friction

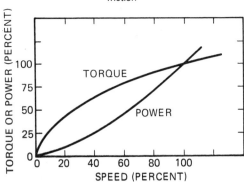

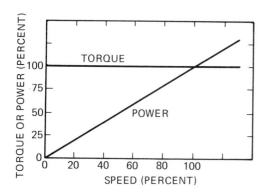

FIGURE 8-3 Torque and horsepower curves for a (theoretical) constant-torque load

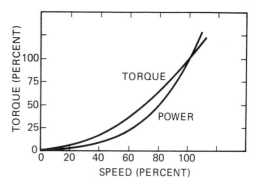

FIGURE 8-4 Torque and horsepower curves for a (theoretical) fan load

type of machine, but two patterns can be recognized:

Constant-Torque Loads

The torque required to drive the machine is essentially constant regardless of speed, and the power required is directly proportional to the speed, as shown in Figure 8-3.

Variable-Torque Loads

The torque required to drive these loads changes from virtually zero at low speed to a considerable value at normal operating speed; the power required rises more rapidly than the torque. Fans, centrifugal pumps, and blowers are common examples. On some types of fans the torque required is proportional to the square of the speed, so the power required is proportional to the cube of the speed, as shown in Figure 8-4.

In practice, fans, blowers, and centrifugal pumps follow their theoretical torque and power curves quite closely, but coulomb friction may give them a small but definite starting-torque requirement (up to 20% of the torque required to drive them at full speed), as shown in Figure 8-5.

Because of viscous friction, most constant-torque loads require less torque at low speeds,

and again coulomb friction tends to raise the starting-torque requirement. In the absence of more accurate information, the curve from Figure 8-5 may be used for such machines. But if the load is cyclic, or if the machine will start unloaded, these factors should be taken into account.

The instantaneous torque needed to drive some loads is not constant but instead is cyclic; that is, it varies regularly according to the instantaneous position of the machine. Punch presses and reciprocating compressors are well-known examples. On still other machines (e.g., rock crushers) the instantaneous driving torque required may be a

FIGURE 8-5 Torque-speed curves for (practical) constant-torque and variable-torque loads

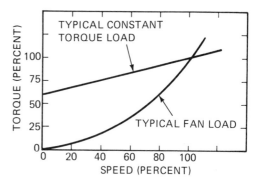

random variable unrelated to machine position. The torque–speed curve of such a machine indicates the average torque needed to maintain a constant average speed, because, when running, cyclic variations of speed[2] will occur permitting inertia to smooth out the torque pulsations considerably. But on starting, inertia is no help, and the starting torque of the motor must exceed the peak instantaneous torque of the load—so cyclic or pulsating loads are hard to start.

For maintenance purposes it must be possible to start and stop the driven machine, and it must also be possible to control the average rate of production or throughput. A constant-speed drive with an on–off mode of operation is commonly used to control production rates; but where the throughput rate is inherently related to machine speed, some form of speed control (either in definite steps or over some continuous range) may be preferable. For some machines it is easiest to control production by directly limiting the flow of processed items or materials, and in such a case the machine is usually driven at some optimum and more or less constant speed, with the throughput independently controlled. When a machine is running but the throughput has been stopped, we say the machine is running unloaded. Unloading generally reduces the torque requirements of the machine (see Figure 8-6) and therefore makes it easier to start and accelerate.

Exact data are scarce, but under loaded

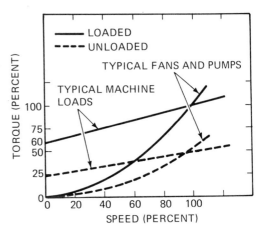

FIGURE 8-6   Effect of unloading on the torque-speed curve

conditions, smooth-running, constant-torque loads can require 125% starting torque, and cyclic constant-torque loads may require 200%. These percentages are based upon the torque required to drive the machine at full speed (loaded) and do not allow any margin for accelerating the load.

Because of the way in which the throughput is controlled, there exists a third category of machines that are classified as constant-power loads. A theoretical constant-power load would require that the driving torque be inversely proportional to the speed; and because operators habitually take heavier cuts at lower speeds, the lathe approaches this theoretical torque curve. The same thing is true of a drill press, where the larger twist drills (requiring more torque) must be run at lower speeds as shown in Figure 8-7. But this constant-power characteristic exists only because of the way the machines are used. With a given drill size and feed rate, a drill press is constant torque in nature.

Before leaving the subject of load devices, we must also consider the effect of reversing the rotation. Many load devices are passive;

---

[2]Cyclic loads always have corresponding cyclic variations of instantaneous speed; otherwise, inertia could have no effect. These instantaneous speed variations can have a significant effect on the driving motor and its power supply system, and we will consider these effects later. The speed variations can also affect the operation of the driven machine, but we will not pursue that train of thought any further.

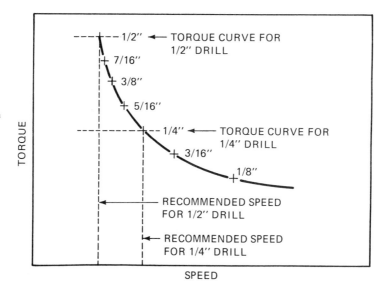

FIGURE 8-7   Torque-speed curves for a drill press

that is, they develop torque only in opposition to their rotation, as shown in Figure 8-8. A horizontal conveyor belt is a typical example.

However, some loads are active; that is, they tend to rotate without having any external driving force applied. A loaded motor-driven chain hoist (except one having non-

reversible worm gearing) is a good example, and a typical torque curve is shown in Figure 8-9. The slight discontinuity in the curve at zero speed is due to coulomb friction.

If an active load tends to accelerate the drive motor above its normal speed, that load is sometimes known as an overhauling load. If the motor then supplies braking torque

FIGURE 8-8   Torque requirement of a passive constant-torque load

FIGURE 8-9   Torque-speed requirement of an active constant-torque load

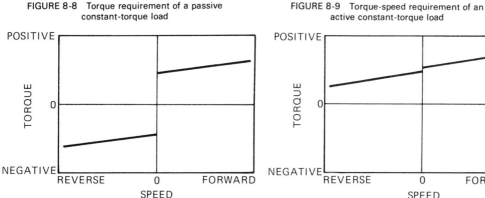

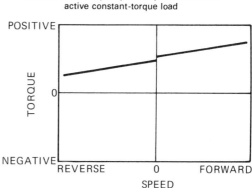

(as most of them will), its power output becomes negative, but this does not harm the motor. To hold an active load at a standstill, a mechanical brake is usually employed.

Choosing a Motor of the Correct Power Rating

The use of a motor that is too small for the application always results in either reduced throughput or reduced motor life, and sometimes both. However, the use of a motor that is larger than necessary will increase initial cost and decrease efficiency, and in the case of most ac motors it decreases power factor as well. For these reasons it is important that the motor be of the correct size for the driven machine.

It is not always easy to predetermine how much power a given machine requires. If the speed and driving torque required at the input shaft are known, the power the load requires can be calculated as follows:

$$P = 2\pi T(\text{rev/s}) \qquad (8\text{--}1)$$

where T = torque in newton-meters (N · m)
    P = power in watts (W)

The speed at which the machine should be driven is usually known or at least readily established. But the torque required to drive a machine is not always known and is not easy to measure in the field. Often the machine manufacturer or supplier will specify the driving power required. Field experience and comparisons with other similar machines may also provide useful guidance. But in the absence of any other reliable information, it may be necessary to do a full-scale test using a temporarily installed motor. Drive the machine at rated speed and at the desired throughput rate, and measure the power input to the motor. If the efficiency of the motor is known or can be determined with

reasonable accuracy, the power required can be calculated from Equation 8–2.

$$\text{Power} = \text{watts input} \times \text{efficiency} \qquad (8\text{--}2)$$

Extreme accuracy is not required because the power requirements of most machines will vary from time to time, and usually the nearest larger preferred standard size of motor will be applied anyway. Remember also that, if the throughput is separately controlled and independent of machine speed, the user will often attempt to raise the production rate after a while, and this will require more driving power.

Unless otherwise specified, the power rating of an electric motor is the maximum power it can develop without exceeding a specified temperature in continuous service, and its full-load torque can be calculated from the power and speed ratings shown on its nameplate, using Equation 8–1. Electrically and mechanically, most motors are capable of developing one and a half to three times their rated power (and torque) without damage provided this condition does not last long enough to overheat the motor. We often take advantage of this fact when applying an electric motor to a cyclic load.

Machines that have cyclic torque requirements generally fall into one of three categories, and the criteria for selecting a drive motor vary accordingly.

MACHINES WITH VERY SHORT CYCLES

On many machines the torque variations are of such a short duration that the inertia of the load will carry it through the peaks, and the drive motor power rating should be based upon the average torque required to turn the load.

MACHINES WITH MEDIUM-LENGTH CYCLES

If the load is periodic in nature (i.e., the

instantaneous power requirement follows a regular cycle), and the cycles are long enough that inertia will not carry the machine through its instantaneous torque peaks and yet short enough that the drive motor will not heat appreciably in one cycle, the power rating of the drive motor must equal the rms power requirement of the load. Root mean square (rms) power is just the square root of the average of the instantaneous power squared. If the power required by the load is (or can be approximated by) a step function, the rms power can be found as illustrated by Figure 8-10 and Equation 8–3.

rms power

$$= \sqrt{\frac{P_1^2 t_1 + P_2^2 t_2 + P_3^2 t_3 + \cdots + P_n^2 t_n}{t_1 + t_2 + t_3 + \cdots + t_n}}$$

$$(8\text{–}3)$$

It should be noted that rms power is always less than the peak but more than the average power requirement, and that Equation 8–3 is valid for cases that involve negative power values as well.

MACHINES WITH LONG CYCLES

If the cycles are long enough that the drive motor will heat appreciably in one cycle, or if

FIGURE 8-10 Power input to a cyclic load

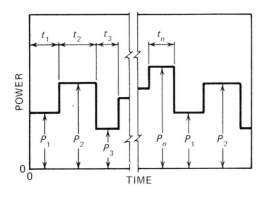

the machine will be started and stopped at regular intervals, the drive motor can still be rated at less than the maximum power required by the load.

However, there is no easy way for the user to determine what size of continuous rated motor to apply in these cases because it depends upon the thermal time constant of the motor. Refer this kind of problem to the motor manufacturer. It should be noted that most manufacturers will supply short time duty (rated) motors on request, with 15-, 30-, and 60-minute ratings being most common. But when seeking their advice, be sure to give them complete information on the duty cycle of the intended application.

If the desired operating speed of the driven machine coincides with the full-load speed of the motor, a direct coupling may be employed; if not, we must connect the motor to its load using belts, chains, or gears selected to obtain the desired drive ratio as defined in Equation 8–4.

Drive ratio

$$= \frac{\text{desired speed of driven load}}{\text{full-load speed of motor}} \quad (8\text{–}4)$$

If the machine does not constitute full load on the motor, the motor speed will usually be slightly high, and so if all factors such as belt slippage, and the like, are carefully allowed for, the machine speed will be slightly higher than desired. But the error involved is usually negligible.

Except for the power loss in the drive train (which is usually small), the drive method and drive ratio have no bearing on the required power rating of the driving motor. The choice of motor speed and drive method are therefore independent of the factors discussed here.

If it is required that the speed of the

driven machine be adjustable, it may be convenient to use a motor that runs at more or less constant speed and drive the load through V-belts and variable-pitch pulleys, or possibly a gear-changing transmission arrangement. In such a case, the motor size must be based upon the highest power demand, which usually occurs at the highest speed. But it may be preferable to use a fixed drive ratio and a motor to which speed control may be applied. If so, the ratio between maximum and minimum motor speeds must be the same as that required by the load (otherwise, some means of changing the drive ratio will still be required), and the continuous duty (power) capability of the motor should match the requirements of the load.

Some adjustable-speed[3] or multispeed[4] electric motors are inherently constant-power machines. Some motors are inherently constant torque,[5] and a few have variable torque capability.[6] If the motor is correctly sized for operation at one speed but the motor's capability at other speeds is not the same as the load's requirement, the motor will either be overloaded at other speeds (a condition that must be avoided) or else lightly loaded (a condition that we prefer to avoid).

It is often advisable to have some margin between the calculated size of motor required and the actual size of the motor installed, particularly if the motor has a service factor[7] of unity. Such a margin will allow for prolonged moderate overload conditions, unusually high ambient temperatures, or unusual voltage conditions.

[3]An adjustable-speed electric motor is a motor that can be operated at any speed within some continuous range.

[4]A multispeed motor is a motor that can be operated at any one of two or more definite speeds (e.g., 29 rev/s or 14 rev/s), but not at any intermediate values.

[5]The continuous duty capabilities of constant-power and constant-torque motors are exactly as their names imply.

[6]The continuous duty torque capability of a variable-torque electric motor is directly proportional to its operating speed, so its horsepower capability is proportional to the square of its operating speed.

[7]The power rating of a motor is based upon a specified temperature rise. If this specified temperature rise does not result in the maximum permissible operating temperature for the insulation in the motor, the motor could carry a continuous overload without damage to itself. The service factor of the motor indicates this degree of permissible overloading; that is, a service factor of 1.15 indicates that a 15% overload is permissible.

ACCELERATION OF THE LOAD

Let us consider a passive load directly coupled to a motor and arbitrarily define forward rotation. Positive motor torque will then be defined as that which tends to cause forward rotation and positive load torque as that which tends to oppose rotation. With these definitions we can then define the net torque acting on the rotating assembly as being the difference between the motor and load torques. Mathematically,

Net torque (T_n) = motor torque (T_m)
$$- \text{load torque } (T_L) \quad (8\text{–}5)$$

By Newton's law, it is this net torque that accelerates the motor and its load. If the net torque is positive, the whole assembly will accelerate in the forward direction (or if it is already turning backward, it will decelerate to zero and then "rev up" in the forward direction), and if the net torque is negative, the acceleration will be in the negative (reverse) direction. Regardless of the actual speed, the acceleration will always be proportional to this net accelerating torque. If we consider the problem in terms of torque–speed graphs, we recognize that the speed can stabilize only at the point(s) where the load torque and motor torque are equal, with the load torque having a greater slope with respect to speed so that the slope of net torque (with respect to speed) is negative. An example of a motor with an adjustable load is shown in Figure 8-11. The torque curve of most motors can be divided into two (and sometimes three) segments:

- The continuous operating range: within which the motor will not exceed its specified temperature rise.
- The momentary range:[8] within which the motor can be operated so long as it is not allowed to overheat.
- An impractical range: some motors have torque capabilities to which they should never be operated because of brush and

---

[8]It is possible to design a motor that can be operated continuously at any point on its torque curve, but such motors are specially designed and built for their application and are found only in small sizes.

commutator problems (or other nonthermal reasons).

With reference to the motor torque curve in Figure 8-11, point S represents starting conditions, point R represents full-load running conditions, and point N represents no-load running conditions. The portion of the curve between point R and the point of equal negative torque $(-R)$ is the permissible range of continuous operation. The portion of the curve between S and R and also that beyond $-R$ represent the momentary capabilities of the motor. This particular motor does not have an impractical range. If the load on the motor follows curve A, the whole assembly will start and accelerate to point Z. If after the motor has reached Z we change the load so that it follows curve B, the motor will drop back to Y and continue to operate at that point. However, if we stop the motor and try to start from rest, it will accelerate only to point X. If the load is changed to follow curve C, the motor speed will always stabilize at point W regardless of its initial

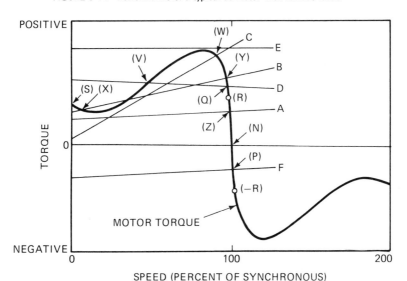

FIGURE 8-11 Performance of a typical ac motor with various loads

speed. With load curve D, the motor cannot start from rest, but if permitted to start unloaded and accelerate beyond point V, it can then run, accelerating to and operating at point O. With load curve E the motor can neither start from rest nor continue to run even with an unloaded start. Curve F is for an overhauling load, and P is the operating point under those conditions.

From an application standpoint there are four things about the motor and load torque curves that should be carefully checked.

Drive Ratio

If the drive ratio is unity and both the motor and load torque curves are expressed in the same unit (*except* percent), their curves can be directly compared. But if the drive ratio is not unity, no such comparison can be made until the load torque curve has been redrawn, multiplying the torque figures by the drive ratio and dividing the speed figures by the same ratio. This new graph is known as a graph of the load torque referred to the motor shaft, and it can be compared to the motor torque graph in the same way as would compare the actual numerical value for a drive ratio of unity.

Net Accelerating Torque

The motor torque capability must exceed the load torque requirements at starting and throughout the speed range up to the intended operating point. A low supply voltage considerably reduces the torque of most motors, and increased friction due to low ambient temperatures can increase the starting torque requirement; so be sure that under normal conditions the load torque requirement is never greater than 80% of the motor torque until the unit is approaching full speed.

Torque Peaks

If the load has a medium or long cyclic characteristic, be sure that the peak load torque does not exceed the maximum (breakdown) torque of the drive motor, because if it does the motor will stall out. If the peak load torque will exceed 80% of the maximum motor torque, it is usually wise to install a larger motor, even though the larger power rating is not warranted by other considerations.

Active Loads

When an electric motor is deenergized, it produces no torque and has little friction. An active load can then accelerate the motor to very high speeds (that is, the motor runs away). For this reason a mechanical brake that is electrically released and spring applied is essential in most of these cases.

EFFECT OF LOAD INERTIA

The inertia of the rotating assembly does not constitute a steady-state load on the drive motor. But inertia does increase the acceleration time and therefore affects the degree of heating that occurs in the motor or its controller during that period. With a squirrel-cage or synchronous motor, the heat energy stored in the rotor during acceleration is approximately equal to the kinetic energy of the motor and its load at full speed, and so this limits the maximum permissible load inertia and/or the permissible number of starts per hour. The same amount of loss occurs during the acceleration of dc or wound-rotor ac motors, but the loss occurs in the accelerating resistors, rather than the motor itself. If the resistors have sufficient thermal storage and/or heat dissipation capabilities, these motors are more suitable for high-inertia loads.

The maximum permissible inertia load for a given motor is usually specified in kilogram-meters squared measured at the motor shaft. If the drive ratio is not unity and the load inertia (at its own drive shaft) is known, the equivalent inertia referred to the motor shaft can be found by taking the actual load inertia and multiplying by the square of the drive ratio.

Some motors and controllers are rated by the manufacturer at so many starts (of specified duration) per hour. The implication is that a larger number of shorter-duration starts (but not vice versa) will be acceptable, and so it then becomes necessary to calculate the acceleration time.

If it is desired to calculate the acceleration time of a given motor and load combination, the necessary equations can be derived from Newton's law as follows: let T_n equal net accelerating torque in pound-feet (lb-ft) referred to the motor shaft[9] and WK^2 equal the inertia of the motor and load in kilogram-meters squared (kg-m^2) referred to the motor shaft.

---

[9]Net accelerating torque (T_n) is the difference between motor torque (T_m) and load torque referred to the motor shaft (T_L). To find T_L, take the actual load torque (at its own shaft) and multiply by the drive ratio.

Rotary acceleration $\dfrac{d\omega}{dt} = \dfrac{T_n}{2\pi WK^2}$ (8-6)

Therefore,

$$dt = \frac{2\pi WK^2}{T_n} \, d\omega \qquad (8\text{-}7)$$

If T_n is constant throughout the accelerating period, we can integrate both sides of Equation 8-7 and get

$$t = \frac{2\pi WK^2\omega}{T_n} \qquad (8\text{-}8)$$

where ω is the change of speed given in revolutions per second and t is elapsed time in seconds. For a motor and load accelerating from rest, ω is the final speed in revolutions per second.

In practice, the net accelerating torque is not constant but instead is a function of speed, and so Equation 8-8 is only an approximation, with the average net accelerating torque being substituted for T_n. They can be applied on a piece-by-piece basis for any motor and load, and reasonably good results can be obtained for squirrel-cage motors using Equation 8-9 and a value for T_n computed in the following manner:

■ Find the average of the starting, breakdown, and full-load torque of the motor.
■ Find the average torque needed to drive the load (i.e., the average from zero to full speed).
■ Take a value for T_n equal to the difference of the preceding steps.

One might be tempted to calculate acceleration time from Equation 8-7 using exact mathematical techniques. But in practice, T_n is not readily available as a mathematical function of ω; even if it is available (or approximated), the resulting equation is difficult if not impossible to integrate.

SPEED REGULATION

Speed regulation is defined to be the change of speed caused by a change from full-load to no-load conditions with all other factors remaining constant. It is usually expressed as a percentage of full-load speed.

In graphical terms, the intersection of the motor torque and load torque curve will always indicate the speed of operation with that particular load. If the load is reduced, the speed will readjust itself to a new point of intersection, and at no load the speed will rise until the motor torque is zero.

Speed regulation is closely related to the slope of the torque–speed curve between full and no load. The more negative the slope, the smaller the speed regulation becomes, as shown in Figure 8-12(a) and (b).

For many applications (e.g., machine tools) it is desirable to have zero or at least comparatively little speed regulation. But high (or poor) speed regulation characteristics can sometimes be advantageous for either of two reasons:

■ A drooping speed characteristic allows the motor to automatically adapt to the requirements of some types of loads. A portable electric drill is a good example (large twist drills should be operated at low speeds.)
■ A drooping speed characteristic relieves the motor of excessive load because of the following—
  □ The effect of speed on the power required by continuous load: if when the load is applied the speed drops to 40% of the no-load value, the power required will be only 60% of that required to maintain a constant speed (assuming a theoretical constant-torque load.)
  □ The effects of inertia on cyclic loads:

a cyclic load, such as a reciprocating air compressor, tends to slow down and speed up as it rotates. If the torque-speed curve of the motor has a high

FIGURE 8-12   Relation between speed regulation and motor torque curve slope

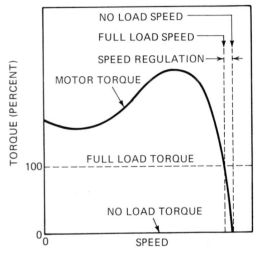

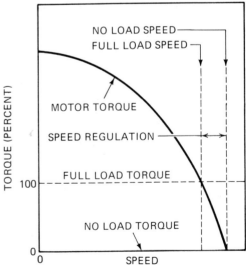

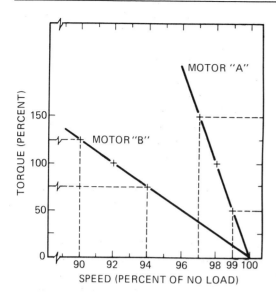

NOTE THAT WITH MOTOR "A", ±1% SPEED VARIATION CAUSES THE TORQUE TO CHANGE BY ±50%. MOTOR "B" HAS A FLATTER TORQUE CURVE AND A CHANGE OF ±2% IN SPEED CAUSES ONLY A ±25% CHANGE IN TORQUE.

(steep) negative slope, the speed variations result in large variations of motor driving torque. The effect is almost as if the motor were trying to force the load through its peak torque requirement and inertia is of little help. But as shown in Figure 8-13, if the torque-speed curve is more nearly horizontal, the motor will permit greater speed variations without developing the same peak torque, and this permits inertia to carry the load through the peaks instead. The result is reduced strain on drive belts and other parts and reduced cyclic variation of power input to the motor. This last advantage is of particular importance because it permits the use of a smaller drive motor (rms power approaches the average) and also reduces the line voltage flutter caused by the cyclic variations of the power input.

FIGURE 8-13 (left) Effect of motor torque curve slope on the torque pulsations with cyclic loads

SPEED CONTROL

The term "speed control" means the ability to operate at various speeds with a given load at the discretion of a human operator or under the control of some external system. Here, again, Newton's law applies, and the motor speed will change until the motor torque and load torque are equal and their net difference has negative slope with respect to speed.

Graphically, speed control is simply a matter of changing the motor torque curve to obtain a new point of intersection with a given load curve, as shown in Figure 8-14(a) and (b). It should be noted in Figure 8-14

that the slope of the torque curve (within the continuous operating range) is not appreciably changed, and so speed regulation is not seriously affected. But some speed-control systems flatten out the torque curve (between the no-load and full-load torque points), as shown in Figure 8-15, and in these cases four interesting things happen.

■ The range of speed control obtained becomes dependent on the shape of the load torque curve. With a given control, a wider range of speed control wll be obtained with a constant-torque load than will be obtained with a fan load.

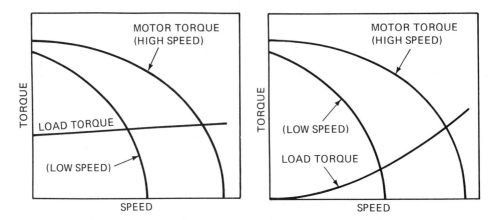

FIGURE 8-14   How changing the motor torque curve provides speed control

FIGURE 8-15   How speed control may affect speed regulation

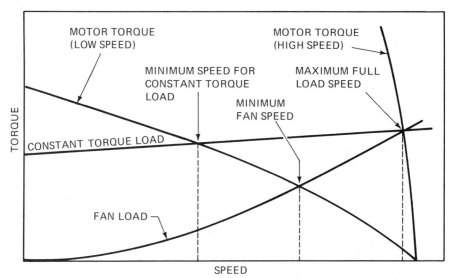

■ When operated at reduced speed, speed regulation of the system will tend to be rather poor.

■ Under no-load conditions, speed control tends to be nonexistent.

■ The action of the speed control will be reversed by an overhauling load; that is, the control changes that reduce speed with a passive load will permit also an increase of speed with an overhauling load.

SUMMARY

Selecting a motor for a given application is done in three broad steps.

■ Determine the requirements of the driven machine in terms of power, torque and inertia, the permissible amount of speed regulation, and the desired range of speed control.

■ Know the capabilities of each type of motor and control in terms of the available range of power ratings, the shape of the torque–speed curve, maximum permissible load inertia, speed regulation characteristics, speed-control capabilities, and overall cost.

■ Select the least expensive motor and control that will meet the requirements of the load.

REVIEW QUESTIONS

8-1. When at rest, many machines exhibit a tendency to stick; that is, the breakaway torque (minimum torque that will cause rotation) is higher than the torque required to turn the machine at low speed. Explain why.

8-2. For each of the machine-and-application cases listed below (in italic, following "(g)"), specify the driving power requirements in the following terms:

(a) Is this a unidirectional or a reversible drive?

(b) Is speed control required; if so—

(i) Is it a variable-torque, constant-torque, or constant-power load?

(ii) Must the speed be adjustable over some continuous range or will two, three, or four definite speeds provide sufficient adjustment?

(c) Is the torque required to turn the machine constant; if not—

(i) Does the torque vary according to the instantaneous position of the machine? or

(ii) Does the torque follow a fairly predictable time-based cycle that is not related to its position? or

(iii) Is the torque a random variable that does not follow any regular pattern?

(d) Does the machine run for a short time period or continuously?

(e) Does it ever become an overhauling load?

(f) Can the machine be conveniently unloaded during the starting period?

(g) Does it have a high-starting-torque requirement?

Grain auger
Passenger elevator
Fan used for—
Ventilation system in an office building
Residential furnace
Portable (desk top) applications
Reciprocating air compressor used in—
Automotive service station

Large factory to operate air-driven tools

Centrifugal water pump

Freight train (to be pulled by an electric locomotive)

Bench-mounted grinder used for—
Home workshop
Grinding rough castings in a high-production factory

Circular saw—
For home workshop use
In a lumber mill, used to cut logs into slabs

Hammer mill used to pulverize coal

Conveyer belt

8-3. Is it correct to say that an electric motor should never be overloaded? Explain.

8-4. A certain load requires 40 kW at 20 rev/s and 30 kW at 15 rev/s. Is this a variable-torque, constant-torque, or constant-power load?

8-5. Explain what is meant by the term "speed regulation."

8-6. What is meant by the expression "drooping speed characteristics"?

8-7. How can one estimate the speed regulation of a given motor by inspecting its torque–speed curve?

8-8. If it is desirable to use a flywheel with a given load device, it may also be advisable to use a drive motor with moderate to poor speed regulation but a definite no-load speed. Explain why.

8-9. Explain what is meant by the term "speed control."

8-10. For use on an electrically driven hoist, the speed-control method must not flatten out the motor torque speed curve. Explain why.

MATHEMATIC PROBLEMS

8-1. If a fan requires 35 kW at 29 rev/s, find (a) the power required at 14.5 rev/s; (b) the torque required at each speed.

8-2. A certain load that rotates continuously goes through a regular 3-min cycle requiring 125 kW for 1 min and 50 kW for the balance of the time. Find the minimum size of drive motor required.

8-3. A load that must run at 29 rev/s is to be driven by a direct-coupled squirrel-cage motor rated at 50 kW. The motor has 150% starting torque and 200% breakdown torque, and its inertia is 5 kg-m^2. If averaged over the range from zero to full speed, the torque required by the load is 140 N · m and the load inertia is 20 kg-m^2. Find the acceleration time.

8-4. The load from problem 8-3 is to be belt driven by a 50-kW 29-rev/s motor that has the same percentage of starting and breakdown torque, but the motor inertia is only 2.0 kg-m^2. If the belt losses are negligible, find the acceleration time.

8-5. The rms power required by a certain load is 25 kW, but its cycle is such that it requires a maximum of 75 kW for intervals of up to 30 s. If the available motors have 200% breakdown torque, what is the minimum recommended size of motor for this application? (Neglect the effect of speed regulation.)

8-6. Under loaded conditions, the torque curve for a certain machine is a linear function of speed, going from 150 N · m of torque at zero speed to 250 N · m at 5 rev/s, which is its normal operating speed. When the machine is unloaded, its torque curve is still linear, from 25 N · m at zero speed to 75 N · m at normal operating speed. The machine will rotate continuously, but its cycle is such that it runs 40 s in the loaded condition, followed by 60 s in the un-loaded condition. The load inertia is 55 kg-m^2 and the motor inertia is 0.2 kg-m^2.

The motor has 115% starting torque and 200% breakdown torque and will have a full-load speed of 29 rev/s. Neglecting belt losses,

(a) Calculate the minimum required power rating of the drive motor.

(b) Calculate the acceleration time for both the loaded and unloaded condition.

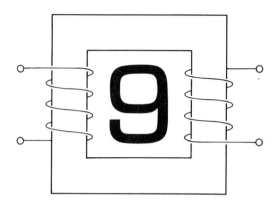

GENERAL INFORMATION ON MOTORS AND ELECTRIC POWER SYSTEMS

Many people who are in nonelectrical occupations need some background on electric motors and electrical power systems. Its not electrical theory they want to know, but rather "what equipment is available and how is it used?" That is what this chapter is intended to provide.

We will start by looking at electrical power systems, with emphasis on what types are available, line drop and voltage flicker, wiring methods, and basic requirements for motor branch circuits. Then we will take a general look at electric motors, with special attention to the nameplate information and

such things as types of motor enclosures. Finally, we will give a brief description of the various types of motors that are available and briefly outline the special characteristics of each.

Although it was originally intended that NEMA standards would be extensively quoted here, the impending metric changeover and the lack of new standards in SI units has forced us to omit most of that information. It seems certain that new standards will be adopted soon and made readily available to industrial concerns and also the general public.

INTERIOR WIRING SYSTEMS

Types of Power Systems

It is essential that electrically operated equipment be suitable for the power supply available at the location where the equipment is to be installed. Assuming that the supply system has sufficient current capacity, the remaining factors are that the frequency

rating of the devices must match that of the power system, and the voltage and phase characteristics of the load and the power system must also be the same. With rare exceptions, the wiring systems in commercial and industrial buildings are alternating current and operate at a frequency of 60 Hz. If a particular device requires direct current, either a motor–generator set or a transformer and rectifier unit will be required to change the available alternating current to the desired value of direct current. Motor–generator sets that will provide a change of frequency are also available, but such conversion equipment is too expensive for general use. So 60-Hz ac equipment is always used unless special circumstances dictate otherwise.

The components of a power system are usually of various manufacture, and so to ensure that their products will be compatible, manufacturers design their equipment for either single or three phase, and have adopted a series of preferred standard voltage ratings for electrical devices.

With a few isolated exceptions, all electrical energy is generated as three phase, and large amounts of power can be most economically transmitted in that form. However, small amounts of power can be more economically distributed over single-phase circuits. For this reason, and because they can easily derive single-phase power from their three-phase system, electrical utilities generally will supply large buildings with three-phase power, while small buildings will often be supplied only with single-phase power. Because three-phase motors are much more reliable and troublefree than single-phase and cost loss, except in very small sizes, three-phase motors are usually preferable to single-phase types. But if the building is not wired for three phase, the cost of installing such wiring may be considerable. If the building is located in an area that is entirely served by a single-phase line (as is often the case in rural or other areas of very low load density), the electrical utility company may charge the cost of the additional lines (needed to bring in three-phase power) to the consumer. The economics of single-phase equipment may then be more

attractive. Converters that will change single-phase to three-phase power are available, but the use of such converters is also limited by economic considerations.

There are five common kinds of power systems that may be installed in a building:

Single Phase

These two types are the simplest power systems and with few exceptions are used only at potentials below 300 volts (V).

TWO-WIRE SYSTEMS   These systems or circuits have only two conductors with a nominal voltage of either 120 or 240 V. For 120-V circuits, it is standard practice to have white insulation on one of the conductors and either red or black insulation on the other. The wires of a 240-V circuit are usually colored red and black. Two-wire circuits are almost always derived from a three-wire system.

THREE-WIRE SYSTEMS   These systems or circuits have three wires, usually colored red, black, and white. The voltage between the colored wires is nominally 240 V, while the voltage between the white and either colored wire is about 120 V. The white wire is known as the neutral. Except for installations in a few older homes, some temporary services on construction sites, and perhaps the *very* smallest of new permanent installations, all single-phase power systems are actually three–wire.

Three Phase

THREE-WIRE SYSTEMS   These systems or circuits have three conductors, usually colored red, black, and blue, and the nominal voltage between any two of the three conductors will be a preferred standard voltage such as 240, 480, 600, or 2400 V. A three-wire three-phase system is satisfactory for motor loads, but since it is not well adapted for lighting equipment and totally unsuitable for 120-V portable equipment, a separate three-wire single-phase system usually takes care of these loads.

FOUR-WIRE WYE SYSTEMS These systems or circuits have four conductors, usually colored red, black, blue, and white, and the voltage between any two of the colored wires is greater than the voltage between any colored wire and the white wire by the factor $\sqrt{3}$. The preferred standard voltages for these systems include 120/208 Y, 277/480 Y,[1] and 346/600 Y[2] volts, where the larger figure is the line-to-line voltage and the smaller one is the line-to-neutral voltage. These systems are suitable for both lighting and motor loads, but if the main wiring system of a building is other than 120/208 Y volts, either a three-wire single-phase or a small capacity 120/208 Y volt three-phase system must be installed to take care of portable appliances.

FOUR-WIRE DELTA SYSTEMS These systems or circuits have four conductors, usually colored red, black, blue, and white, and the voltages between any two of the colored wires are equal. However, the voltage from one of the colored wires to the white neutral is about 86% of the line-to-line voltage, while the voltage from either of the remaining colored wires to the white is about 50% of the line-to-line voltage. This system is used only at or about the 240-V level, and is intended to supply combined motor, lighting, and portable appliance loads. The popularity of this system is declining. It amounts to a three-wire three-phase and a three-wire single-phase system combined into one.

[1]Preferred in the United States but not in Canada.
[2]Preferred in Canada but not in the United States.

One-Line Diagrams

The most explicit representation of a wiring system is probably a complete wiring diagram that shows each conductor in each phase of the circuit. However, for short-circuit studies and for planning a wiring system, such detail is not necessary and tends to be confusing; so we use the one-line diagram for these tasks. Many of the symbols used are the same or a least similar to the symbols used in a wiring diagram, but the diagram looks almost as if only one circuit conductor were shown. Figure 9-1 is a typical example.

Note that no effort is made to show any control circuits, or any details of the wiring. Notice too that short notes may be included on the diagram to help relate the symbols to the proper components of the actual system.

The advantage of the one-line diagram is that it shows very clearly which components of the wiring system carry current to the loads on any particular branch circuit.

Line Drop and Voltage Regulation

Line drop is the voltage required to circulate current through the connecting wires of a circuit, and it is equal to the product of

FIGURE 9-1 Typical one-line diagram

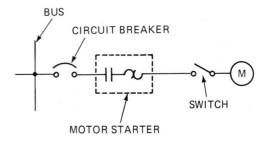

load current and line resistance. It has the effect of reducing the voltage of the load below that at the source. A steady-state low-voltage condition is objectionable for two reasons. First, it reduces the starting torque developed by a motor and will cause it to overheat when driving its normal load. On ac motors, starting torque is proportional to the square of the voltage. Second, low voltage reduces the output of lighting and heating equipment.

It is possible to compensate for line drop by slightly raising the voltage at the source.[3] However, overvoltage shortens the life of lamps and heating equipment and can also cause motors to overheat. So the source voltage must not be raised too much, because its increased value will be impressed upon nearby loads at all times, and if most of the load is switched off, it will appear throughout the system.

Another important aspect of line drop is light flicker caused by motor starting currents.[4] Consider a transformer, feeder, and branch circuit arrangement as shown in Figure 9-2. For economic reasons, transformer ratings and conductor sizes are based upon the full-load currents of all the loads they supply. The total voltage drop in the transformer and feeder conductors of such a system might be 2% or so at full load, with additional voltage drop occurring in the branch circuit conductors to the motors. However, if a certain motor on such a system constituted 16% of the total load, its starting current would be about equal to the full-load current on the system, and so the voltage drop caused by its starting current

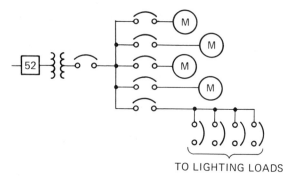

FIGURE 9-2   Transformer, feeder, and branch circuit arrangement

would be about 2%. The larger the motor (as a percentage of supply system capacity), the greater the voltage disturbance it can create. Furthermore, the fact that the impedance of most supply systems is predominantly inductive reactance, coupled with the fact that motor starting currents usually lag the voltage by a large angle, makes the voltage dip caused by motor starting worse than would be expected just by looking at the voltage drop that occurs at full load.

The voltage drops due to the impedance of the transformers and the secondary feeders are most important because these are observable at all loads on the low-voltage system. Primary line drop is usually negligible, and while the motor is affected by line drop in its own branch circuit, other loads are not. Except for very sensitive devices such as computers, the voltage dips caused by motor starting are rarely severe enough to affect the operation of other machines on the system, and people would not become aware of the voltage-drop problem except for one thing: the dips can cause the lights on the system to flicker each time a motor is started. This light flicker is sometimes most annoying, and it is from this phenomenon that most complaints originate.

---

[3]Most transformers have taps so that this can easily be done.

[4]These are short time variations and are often called voltage flicker.

Light flicker can also be caused by cyclic loads such as reciprocating compressors, but this problem is not common.

Several relevant factors should be kept in mind:

- High-capacity systems can carry the starting current of large motors without distress. In large industrial plants it may be quite practical to start 2000-kilowatt (kW) motors across the line, while in commercial or light industrial areas a 25-kW motor may cause objectionable voltage dips.
- The sensitivity of different individuals will vary considerably, but on the average light flicker becomes noticeable if a sudden voltage dip of about 1.5% (range 1 to 2.5%) occurs on 100-watt (W) incandescent lamps or if about 4% dip (range of 2.5 to 5.5%) occurs in 1000-W lamps. Exact data are scarce, but it is known that gradual voltage dips are not as noticeable. It appears that voltage flicker is not as readily observed on fluorescent or mercury-vapor lighting as it is on incandescent.
- An occasional voltage dip (e.g., caused by starting a motor once every 3 or 4 hours) might be tolerable, whereas frequent voltage dips (caused by repetitive starting of an automatically controlled motor) would be objectionable.
- Light flicker is especially annoying to people who are reading, writing, or typing. Voltage dips that go unnoticed in an industrial plant may be objectionable in an office or school.
- Consumers will often object to voltage dips of unknown origin when they will tolerate voltage fluctuations caused by operation of their own equipment. If a power company uses one transformer bank to feed several consumers, it may

be necessary for the company to establish rules restricting the motor-starting current that any one consumer may draw from the line.

A third important result of line drop is voltage regulation, which is the change of voltage due to a change of load current. Although they are essentially the same phenomenon, transient voltage dips caused by motor starting currents are specifically excluded from this discussion. The electrical utilities company usually tries to maintain a constant voltage at or near its own substations. However, since the load current is not constant but varies in a cyclic pattern during a 24-hour day and also varies with the seasons, it is obvious that line drop will cause the voltage at the consumers' premises to vary from time to time, depending upon how much load is in use. The voltage drop along the interior wiring of the building also tends to cause low voltage at the load and poor voltage regulation (i.e., considerable voltage variation.) Because additional loads are usually added from time to time, this problem is commonly found in older buildings.

The Canadian Standards Association has established acceptable limits of voltage variations at the point where the electrical power enters the building (service entrance) and also at the load terminals under both normal and extreme operating conditions. These limits are shown in Tables 9-1 and 9-2.

Prolonged voltage excursions into the extreme operating range indicate the need for corrective action. Prolonged voltage excursions beyond the extreme operating range may result in damaged equipment or forced shutdown.

If the problem is drawn to their attention, most electrical utility companies will correct

TABLE 9-1   VOLTAGE VARIATION LIMITS AT THE LOAD TERMINALS

Nominal System Voltages	Voltage Variation Limits Applicable at Utilization Points			
	Extreme Operating Conditions — Normal Operating Conditions			
Single phase				
120/240	104/208	108/216	125/250	127/254
240	208	216	250	254
480	416	432	500	508
600	520	540	625	635
Three phase four conductor				
120/208Y	108/187	110/190	125/216	127/220
240/416Y	216/374	220/380	250/432	254/440
277/480Y	240/416	250/432	288/500	293/508
347/600Y	300/520	312/540	360/625	367/635
Three phase three conductor				
240	208	216	250	254
480	416	432	500	508
600	520	540	625	635

With the permission of CSA, this material is reproduced from CSA Standard C 235-1969 Preferred Voltage Levels for AC Systems, 0 to 50 000 V which is copyrighted by the Canadian Standards Association, and copies of which may be purchased from the Association, 178 Rexdale Boulevard, Rexdale, Ontario M9W 1R3, Canada.

the voltage supplied to the consumers' service. To keep the voltage drop within acceptable limits, it may be necessary to use oversized conductors on long runs of interior wiring.[5] If voltage drop is a serious problem on an existing installation, extensive rewiring is probably necessary.

Electrical Fire Hazard

Fires that have an electrical origin get started in one of three ways:

Short Circuits

A short circuit (which includes ground faults on grounded equipment supplied from

[5]See *The Canadian Electrical Code, Part I*, 12th ed., Table D3 (Rexdale, Ont.: The Canadian Standards Association, 1975).

a grounded system) results in a very high current flow, ten to several hundred times the normal value. Such a high current will cause rapid heating of the circuit conductors, and the insulation on the wires will almost certainly catch fire if the current is not quickly interrupted. Considerable heating and probably some arcing and melting of the circuit conductors will also occur at the point of the short circuit, and this too may cause a fire.

Overloading

This raises the temperature of the conductors to an abnormal level, and while it is not likely to be the direct cause of a fire, it causes accelerated deterioration of the insulation and may result in a short circuit.

TABLE 9-2   VOLTAGE VARIATION LIMITS AT SERVICE ENTRANCES

Nominal System Voltages	Voltage Variation Limits Applicable at Service Entrances			
	Extreme Operating Conditions			
	Normal Operating Conditions			
Single phase				
120/240	106/212	110/220	125/250	127/254
240	212	220	250	254
480	424	440	500	508
600	530	550	625	635
Three phase four conductor				
120/208Y	110/190	112/194	125/216	127/220
240/416Y	220/380	224/388	250/432	254/440
277/480Y	245/424	254/440	288/500	293/508
347/600Y	306/530	318/550	360/625	367/635
Three phase three conductor				
240	212	220	250	254
480	424	440	500	508
600	530	550	625	635

With the permission of CSA, this material is reproduced from CSA Standard C 235-1969 Preferred Voltage Levels for AC Systems, 0 to 50 000 V which is copyrighted by the Canadian Standards Association, and copies of which may be purchased from the Association, 178 Rexdale Boulevard, Rexdale, Ontario M9W 1R3, Canada.

Poor Electrical Connections

The flow of normal current through a poorly made splice results in heating at that point and can cause the insulation to burst into flame.

There are three ways in which the fire hazard can be reduced. One technique is to protect the current-carrying parts (including the wires themselves) from any kind of mechanical injury that might result in a short circuit. Another is to enclose all current-carrying parts so that if trouble develops the damage will be confined to the space within the enclosure. The third important technique is to protect the system against overload and overcurrent (short circuit) conditions by means of properly chosen and installed fuses, circuit breakers, and/or overload relays.

Equipment Enclosures

The boxes used to enclose electrical devices serve three purposes:

■ They protect personnel from contact with live or moving parts.
■ They protect electrical devices from foreign contaminants and/or mechanical injury.
■ If a fault occurs, they confine the damage to the space within the enclosure.

Almost any kind of enclosure will serve the first purpose, but the type of enclosure needed to perform the other two functions depends upon ambient conditions. The *Canadian Electrical Code* recognizes the following types of enclosures[6]:

[6]See *The Canadian Electrical Code, Part I*, 12th ed., Rules 2-400 and 18-012.

- *Enclosure 1:* a general-purpose enclosure of metal or other suitable material for use indoors in ordinary locations.
- *Enclosure 2:* (driptight) an enclosure designed to exclude falling moisture or dirt for use indoors where the equipment may be subject to falling liquids due to condensation.
- *Enclosure 3:* (weatherproof) an enclosure designed for use outdoors where it may be subject to weather, falling moisture, or external splashing.
- *Enclosure 4:* (watertight) an enclosure designed so that a stream of water from a hose will not enter the enclosure.
- *Enclosure 5:* (dusttight) an enclosure designed so that dust, ignitable fibers, or combustible flyings cannot enter the enclosure.

Special Types for Hazardous Locations

For areas where combustible or electrically conductive dusts are present, enclosures specially designed for the application must be used. These are sealed up somewhat better than a type 5 enclosure.

In areas where explosive or inflammable gases or vapors are present, it is not possible to keep the vapors out of the electrical enclosures. An enclosure designed for such locations is therefore built strong enough to withstand the force of any explosion that can (and periodically will) occur within it, and well sealed to prevent the hot gases from escaping and causing an external explosion.

Wiring Methods

The expression "wiring methods" refers to the way in which the circuit conductors are run from place to place. There are only three forms that the circuit conductors can take and only a few basic manners in which they are installed.

Insulated Wires

These consist of a single metallic path (which may be either solid or stranded to provide moderate flexibility) covered with insulation such as rubber or cross-linked polyethylene. Insulated wire intended for use at up to 600 V is readily available in sizes rated at 15 to 500 amperes (A).

When used for building wiring, insulated wire must always be installed in some kind of an enclosure. This is most commonly done by installing a system of conduit or tubing in the building during its early stages of construction, and pulling or fishing the wires into the conduit after the wall and ceiling finishes have been applied.

Buses

A bus is an aluminum or copper conductor, usually of solid rectangular cross section, that is supported on rigid insulators. Short lengths of buses are used where a feeder divides into a number of separate circuits, but to carry currents in the range of 225 to 4000 A, bus duct (an assembly of buses mounted in a sheet-metal enclosure) is commonly used.

Bus duct intended for use at up to 600 V is available with either three or four conductors for use in either three- or four-wire three-phase systems. Manufacturers supply it in straight sections up to about 3 meters (m) long, and also elbows, bends, offsets, and the like; this type of wiring is ordered and custom made for the building in which it is installed.

Some types of bus duct are a plug-in

type; that is, the manufacturer will supply fused switches or circuit breakers that can be plugged into the bus duct at regular intervals of about 0.6 m along its length. This makes it easy to supply power to any loads that may be added after the building is completed. But some types of bus duct (known as low-reactance or low-impedance duct or paired phase feeder duct) have insulated bars and cannot be tapped in this manner.

Prefabricated Cable Assemblies

These consist of two or more insulated wires, usually enclosed in a common outer jacket. For building wiring, paper-insulated lead-covered cable (PILC) is popular for high-voltage circuits, mineral-insulated cable is used at up to 600 V, and aluminum-sheathed cable can be used at up to 300 V. Cabtire cable (commonly known as extension cord) is mainly intended for portable equipment and must not be used as a substitute for the fixed wiring of a building.

Basic Requirements for Motor Branch Circuits

A branch circuit is that portion of the building wiring that extends beyond the final set of overcurrent devices. An overcurrent device is one that will automatically open a circuit under both overload and short-circuit conditions. A circuit breaker or a fuse fits this definition.

An overload device is a device that will automatically open a circuit if the current is moderately above normal (up to about 10 times rated current), but it is not designed to interrupt short-circuit currents, which are ordinarily much higher.

For circuits that supply lighting loads, the fuses or circuit breakers provide both overload and overcurrent protection. However, for motor circuits, fuses or breakers usually provide overcurrent protection only, and a separate device known as an overload relay provides overload protection for the motor and also the wiring.

Because of their inductive nature, motors are difficult loads to switch. The induced voltage surge produced when interrupting the motor current is particularly troublesome because it causes severe arcing at the switch contacts. A switch that is intended to be used to start and stop a motor must be specifically designed for the purpose and is called a motor starter.

A disconnecting means is a device that can disconnect the conductors of a circuit from their source of supply.

Whether single or three phase, the branch circuit supplying a motor must always be supplied through a manually operable disconnecting means and a set of overcurrent device(s), and it must include a motor starting switch and overload device(s), all of which must appear in that order as we move from the supply source toward the motor. The overcurrent devices are usually combined with the disconnecting means (a fused switch or a circuit breaker), and the overload devices are usually built into the motor starter; thus the wiring layout will be as shown in Figure 9-3.

The fused switch must be located at or very near the point at which the conductor sizes are decreased, and it defines the start of the branch circuit. There must be a fuse (tripping mechanism) in each ungrounded conductor, and if operated manually, the switch (or breaker) must simultaneously open all ungrounded conductors in the circuit.

If the motor starter is a manually operated

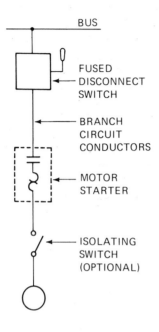

BUS

FUSED
DISCONNECT
SWITCH

BRANCH
CIRCUIT
CONDUCTORS

MOTOR
STARTER

ISOLATING
SWITCH
(OPTIONAL)

type, it is usually located on the unit it controls or else on the wall nearby; if it is a magnetically operated device, it may be located immediately on the load side of the fused switch (a motor control center type of arrangement) some distance from the motor.

For the safety and convenience of maintenance personnel, it is advisable to have some means of switching off the power supply to a motor from a point within sight of the motor. The fused switch for the circuit or a manual motor starter may serve this purpose if suitably located, but in other cases an unfused switch may be mounted near the motor.

FIGURE 9-3 (left)   Motor branch circuit

ELECTRIC MOTORS

Types of Motors

The more common types of electric motors that are manufactured and sold as individual devices are shown in Figure 9-4. The names applied to various groups of motors are shown thereon enclosed in square brackets; where a given motor is commonly known by different names, the alternate names are included in round brackets. Some seldom encountered types have been omitted here.

Interpretation of Nameplate Data

ESSENTIAL NAMEPLATE INFORMATION   Rule 2-100 of *The Canadian Electrical Code**

*With the permission of CSA this material is reproduced from CSA Standard C 22.1—1978, The Canadian Electrical Code, Part I, 13th Edition, which is copyrighted by the Canadian Standards Association, and copies of which may be purchased from the Association, 178 Rexdale Boulevard, Rexdale, Ontario M9W 1R3, Canada.

requires that:

Each piece of electrical equipment shall bear such of the following markings as may be necessary to identify the equipment and ensure that it is suitable for the particular installation:

a. *the maker's name, trademark, or other recognized symbol of identification;*
b. *catalogue number or type;*
c. *voltage;*
d. *rated load amperes;*
e. *watts, volt amperes, or horsepower;*
f. *whether for AC, DC, or both;*
g. *number of phases;*
h. *frequency in hertz;*
i. *rated load speed in revolutions per minute;*
j. *designation of terminals;*
k. *whether for continuous or intermittent duty;*

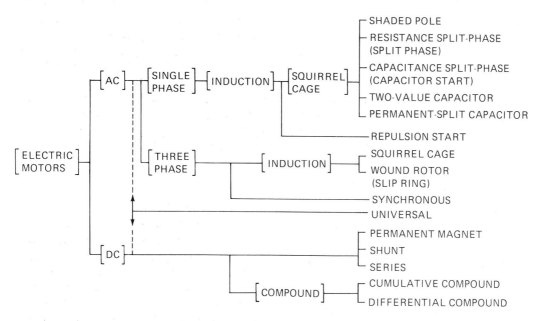

FIGURE 9-4   Classification of electric motors

l.   evidence of approval; and

m.   such other markings as may be necessary to ensure safe and proper operation.

Let us examine each of the above items with reference to motors.

■ *Item a:* Self-explanatory.

■ *Item b:* Refer to the manufacturer's catalogues for information.

■ *Item c:* The voltage marked on the nameplate is that on which all other data are based. Most motors will operate satisfactorily if the applied voltage is within ±10% of their nameplate rating (other factors remaining at rated values.) See Table 9-1. Prior to about 1970, motors were rated at 220, 440, or 550 V for use on 240-, 480-, or 600-V systems, but 220-V motors were also usable on 208-V systems. Current practice on ac motors is to supply them in 200-, 230-, 460-, or 575-V ratings for use on 208-, 240-, 480-, or 600-V systems, respectively. It must be emphasized that present-day 230-V motors are not suitable for use in 208-V systems. Motors marked 230/460 V can be connected for use at either voltage (i.e., when properly connected, they develop the same torque and power at either voltage). Operating a motor below rated voltage lowers the motor's torque curve and increases the motor current at full load so that the motor overheats. Operating above rated voltage causes ac motors to draw excessive magnetizing current, thus causing overheating even at no load.

■ *Item d:* The current marked on the motor nameplate is the current it draws with rated voltage and full load. At no load the current will be less than the marked value, and the starting current for an ac motor will be about six times the rated value. If more than one current is quoted on the nameplate, the motor is either a multivoltage, multispeed, or multipower design with the currents listed in the same order as the voltage (or speed or power) ratings. For example, 7.5 kW 230/460 volt, 28/14 amperes. This motor will develop 7.5-kW output and when so doing will draw 28 A from a 230-V system or 14 A with 460 V applied. Measuring the current to a newly installed motor is the most valuable check one can perform. If the current does not exceed the motor rating, the installation will usually be satisfactory.

■ *Item e:* Unless otherwise indicated, the power rating of a motor[7] is the maximum permissible continuous output that it can maintain without exceeding a specified temperature rise. Increasing the load on a motor causes it to draw more current, and this will cause its temperature rise to increase; but most motors will develop far more than their rated power for short time periods, and short-time overloading is usually permissible as long as the motor does not overheat. Motors have been rated in horsepower for many years (1 hp = 746 W), but this practice will soon be discontinued in favor of the kilowatt.

■ *Item f:* Self-explanatory.

■ *Item g:* Self-explanatory.

■ *Item h:* Most ac motors are designed to withstand a ±5% variation of frequency (other factors remaining at rated values), but except for emergencies such frequency excursions are unthinkable on a utility system. A few motors have been designed to operate at the same voltage and at either 50 or 60 Hz without any change. Raising the frequency of the supply will increase the speed of an ac motor in direct

proportion, but motor torque varies inversely as the square of the frequency.

■ *Item i:* The speed marked on the nameplate of a squirrel-cage motor is its full-load speed. On wound-rotor machines the maximum full-load speed is indicated. However, because its speed is adjustable, the speed marked on a dc machine tends to be somewhat nominal (neither base speed nor maximum obtainable). On older machines the speed is likely to be given in revolutions per minute (rpm), but that term is not part of the SI system of units. We have used the term revolutions per second (rev/s) in this book.

■ *Item j:* On dc or single-phase ac motors that have only two external terminals, or three-phase motors having only three external leads, the manufacturer usually does not mark the terminals. But if any confusion is possible, the leads must be marked and a diagram provided so that the installer can connect the machine without difficulty. The diagrams in Chapters 2, 4, 5, and 6 cover the most common types.

■ *Item k:* Self-explanatory.

■ *Item l:* This refers to approval by the Canadian Standards Association (CSA approval). In most parts of Canada the electrical inspection authorities require that every piece of electrical equipment be approved by CSA. Underwriters' Laboratories (UL) approval carries similar authority in the United States but is not recognized in Canada.

■ *Item m:* One of the most common points to be included here is the type of motor enclosure, particularly if the motor is intended to operate in atmospheres containing flammable vapors (or gases or dusts), conductive dusts, or easily ignitable fibers or flyings. For details, see Section 18 of the *Canadian Electrical Code,* Part I

[7]This definition does not apply to portable tools or household appliances.

(CSA Standard C 22.1, 1975). Typical quotations from motor nameplates are:

Approved for Class I, Division I, Groups F and G

Enclosure: DP (drip proof)

ENCL: TEFC (totally enclosed fan cooled)

ENCL: TENV (totally enclosed not ventilated)

ADDITIONAL NAMEPLATE INFORMATION All or any of the following items may appear on the nameplate of a motor.

1. *Insulation class:* The types of insulation used for motor windings are classified in terms of the maximum temperature that these materials can withstand continuously with reasonable life expectancy. The classes and temperature limits are as given in Figure 1–7 (see page 6). The latest models of standard motors (T-frames) are wound with class B insulation. Older machines are generally class A.

2. *Temperature rise:* The difference between the temperature of a motor and the temperature of the surrounding air is known as the temperature rise of the motor. Manufacturers have adopted standards of maximum temperature rise (such as 40, 55, and 65°C, based upon a 40°C ambient and full load), and this standard is usually shown on the motor nameplate. It is obvious that the operating temperature of a motor is directly related to ambient temperature. If a motor is placed in an area where the temperature is high, it may exceed the temperature rating of its insulation and therefore have a very short life. In such a case, rewinding the motor with insulation having a higher temperature rating is often the best solution.

3. *Service factor:* If the actual full-load temperature rise of a motor plus the ambient temperature is less than the temperature rating of the insulation, it is obvious that the motor could carry some degree of overload continuously without appreciably reducing its life. The term "service factor" indicates the degree of permissible continuous overload with other factors remaining at rated values. For example,

Service factor 1.0 (overloads are not permitted)

Service factor 1.15 (15% overload is permissible)

4. *Frame size:* The National Electrical Manufacturers' Association (NEMA) has adopted a series of preferred standard dimensions for electric motors. Each standard size is designated by a number known as a frame number, and the series extends to about 200 kW at 30 rev/s. Motors that have the same frame number have the same[8] critical dimensions, and therefore (if they have the same power, voltage, and speed ratings) they are interchangeable. The dimensions and tolerances of these frame sizes are available from almost any manufacturer.

5. *Code letters:* The code letter has been used to indicate the numerical ratio between kVA input under locked rotor (starting) conditions and the full-load horsepower rating of the motor. The numerical values of the code letters are given in Table 9-3. This table will probably be revised when new standards are developed in SI units.

Because of Underwriters' Laboratories requirements, the code letter almost always appears on equipment sold in the United States, and as a result it may appear on American-made equipment sold in Canada. But CSA does not require any code letter markings, so it is not always shown on motors destined for Canadian markets.

[8]The word "same" must be interpreted with due regard for manufacturing tolerances. For example, if a direct-coupled motor is replaced, realignment of the coupling will be required.

TABLE 9-3   NUMERICAL VALUES OF CODE LETTERS

Code Letter	Starting kVA per Hp
A	0–3.14
B	3.15–3.54
C	3.55–3.99
D	4.0–4.49
E	4.5–4.99
F	5.0–5.59
G	5.6–6.29
H	6.3–7.09
J	7.1–7.99
K	8.0–8.99
L	9.0–9.99
M	10.0–11.19
N	11.2–12.49
P	12.5–13.99
R	14.0–15.99
S	16.0–17.99
T	18.0–19.99
U	20.0–22.39
V	22.4–and up

Reprinted by permission from NFPA 70-1978, National Electrical Code®, Copyright © 1977, National Fire Protection Association, Boston, MA.

6. *Power factor:* European motor manufacturers frequently mark the full-load power factor of a squirrel-cage motor on its nameplate. A typical nameplate quotation is "Cos ϕ 0.82," but this practice is not followed in North America. However, the minimum permissible continuous power factor (leading) at which a synchronous motor can carry full load is always shown on the nameplate of such motors.

Motor Enclosures

Large electric motors are frequently of open construction (ventilation openings on all sides, suitable for clean, dry, indoor locations only), but dripproof construction (ventilation openings only on the underside) is more common in sizes below 500 kW.

For locations requiring weatherproof, watertight, or dusttight equipment, totally enclosed motors are available. In 5-kW sizes and up, these usually have a fan arranged to blow air over the outside of the case to help cool the motor (called a totally enclosed fan-cooled motor, abbreviated TEFC). The fan is usually omitted on totally enclosed motors below 5 kW, and they are then known as totally enclosed not ventilated (TENV) motors. Note that outside air does *not* circulate around the winding of any totally enclosed motor.

Explosion-proof motors suitable for use in most kinds of hazardous locations are also available. They look much like TEFC or TENV motors, but the case or enclosure will be thicker and stronger to withstand the force of an internal explosion.

Protection from Moisture

The temperature rise of a motor inherently protects it from condensation while in service. But when the motor is shut down, it cools to ambient temperature, and moisture may condense on the windings, particularly if the motor is located in a cold, damp area. To prevent this on a large motor, one can usually install heating elements within the enclosure and wire them through normally closed contacts on the motor starter. Heating will then be automatically applied whenever the motor is deenergized. In the case of small motors, there may not be enough space in which to mount the heaters, but the same effect can be obtained by applying a low voltage to the motor windings.

Connecting a Motor to Its Load

The full-load torque developed by a motor is roughly proportional to its physical size (peripheral area of the rotor). Since power is proportional to the product of torque and speed, a 10-kW 30-rev/s motor, a 7½-kW 20-rev/s motor, or a 5-kW 15-rev/s model will all be about the same size. Since the cost of a motor is roughly proportional to its physical size, all the above motors will be about the same price. Stated another way, low-speed motors tend to be bulky and expensive, and, for reasons that will not be explained here, they tend to be inefficient. In the case of ac induction motors, they have lower power factors.

At first glance, the most economical and space-conserving way of connecting a motor to its load appears to be a direct coupling; but induction motors are available with only certain full-load speeds, and unless the desired load speed corresponds to one of the available motor speeds, direct coupling is impossible. Furthermore, the lower cost of a high-speed motor may make it economical to use such a motor and a speed-reducing drive, even though a motor suitable for direct coupling is available.

If the motor speed and the desired load speed are not the same, the mechanical connection between them will have to be made with either belts, chains, or gears. Many factors are involved in selecting a suitable drive connection, but, basically, chain drives tend to be impractical at high speeds and belt drives tend to be impractical at very low speeds. If a high-speed motor is to be used to drive a very low speed machine, a gear-type speed reducer is often the best answer.

A gear motor is just a reduction gear assembly with a motor attached. A large range of power ratings and output speeds down to about 1.6 rev/s is readily available.

When a motor is used on belt or chain drive arrangements, it should be mounted on a sliding base to facilitate belt tension (or chain wear) adjustments. But care must be taken to avoid the following, which will overload the drive end bearing.

■ Too much belt tension.
■ A pulley that is too wide; this puts the belt tension to far outboard of the drive end bearing.
■ A pulley of too small a diameter; this will necessitate more belt tension to prevent slippage. Reducing the pulley diameter also increases the bearing load because of increased differential tension on the two sides of the belt or chain loop.

Consult the motor manufacturer or the relevant NEMA standards for minimum pulley diameters and maximum widths, and remember that some two-pole motors are not recommended for belt drive.

Mounting Position and Mounting Provision

Most electric motors are designed and assembled for mounting on a horizontal surface, with the shaft in a horizontal position. For sidewall or ceiling mounting with the shaft horizontal, it is usually possible to rotate the end shields so that the ventilation openings are in the desired position (usually down), and, if it is desired, one can usually reverse the stator relative to the shaft tension so that the connection box is located on the other side of the motor. If mounted with the shaft vertical, the force of gravity on the rotor can cause excessive bearing wear, particularly in large machines or sleeve-bearing types. For vertical shaft applications, consult the motor manufacturer, because many motors can be equipped with thrust-type bearings if required. Specially designed vertical shaft motors are also available.

To minimize noise, single-phase motors are often mounted by means of rubber end rings clamped to a cradle base. This arrange-ment is also known as a resilient base. How-ever, dc and three-phase motors are inher-ently quieter than single-phase types. As a result, they normally have mounting feet that are an integral part of the motor frame.

CHARACTERISTICS OF DIRECT-CURRENT MOTORS

Construction and Operation of DC Motors

The heart of a dc motor is the armature, that is, the rotating assembly consisting of the shaft, the laminated core, the armature winding (which lies in slots in the core), and the commutator (to which the armature coils are connected.) The power supply is carried to the armature by means of the brushes, which rest on the commutator. But in order to function, an armature must be placed in an external magnetic field, and so we commonly classify dc motors in terms of the way in which this magnetic field is provided.

When power is first applied to a dc motor, the armature current tends to be very high, 8 to 12 times the full-load current. As the motor accelerates, it generates a counter emf in the armature winding, and this reduces the current and torque. But except for motors of less than about $\frac{1}{4}$ kW, the starting cur-rent must be limited to perhaps twice the full-load current or else cumulative burning and erosion of the brushes and commutator will occur. So we ordinarily start a dc motor with resistance in series with the armature, and bypass this starting resistance when the motor has accelerated sufficiently.

Direct-current motors are available in sizes ranging from a few watts to several thousand kilowatts. But because they are very expensive and require more maintenance than ac machines, dc motors are not com-monly used except on applications where speed control is required.

If the starting current is held to moderate values, heating during acceleration occurs mainly in the starting resistors. If the start-ing resistors are appropriate, dc motors can accelerate high-inertia loads without distress.

Permanent Magnet Motors

As the name implies, these motors have permanent magnets bolted to the field frame and do not require any field coils. If a perma-nent magnet motor is equipped with field coils, those coils are intended for recharging the magnets only.

The torque curve of a permanent magnet motor is a straight line, as shown in Figure 9-5. Theoretically, this motor can develop extremely high locked rotor torque,[9] but if we limit the armature current to 200%, the maxi-mum torque will also be 200% of the full-load value. These motors normally have good speed regulation, 10% or less.

There are two ways of controlling the speed of these motors:

■ Use a constant-voltage power supply and connect resistance in series with the motor

[9]Some manufacturers claim 700% starting torque for these motors.

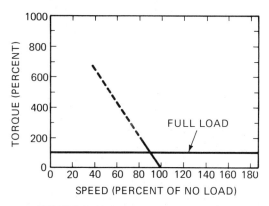

FIGURE 9-5 Typical torque–speed curve for permanent magnet motor

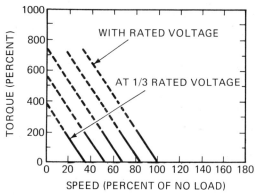

FIGURE 9-7 Effect of voltage on the torque–speed curve

armature. With this method, the torque curve is flattened, with the no-load speed remaining the same as shown in Figure 9-6. At reduced speeds this method of control causes poor speed regulation, and the power losses in the resistor cause poor efficiency. (Overall efficiency is proportional to motor speed.)

■ Use an adjustable voltage power supply. With this method, the torque curves at various voltages remain geometrically parallel, with the no-load speed being directly proportional to the voltage, as shown in

FIGURE 9-6 Effect of armature resistance on the torque–speed curve

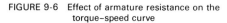

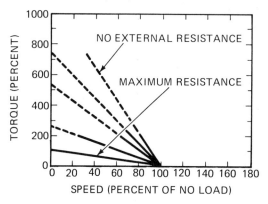

Figure 9-7. Total losses (in watts) and speed regulation remain the same at all speeds.

With either method of control, the losses within the motor are a function of the torque but independent of speed. The maximum permissible continuous torque that the motor can develop without overheating therefore tends to be the same at any speed, although additional ventilation may be required for low-speed operation.

Shunt-Wound Motors

The field flux in these motors is produced by a set of field coils that are connected across the line. The field flux therefore tends to be constant, as it is in permanent magnet machines, giving the same motor performance and the same possibilities for speed control.[10] But it is possible to connect a rheostat in series with the shunt field coils and thereby obtain another form of speed control;

[10]For the armature resistance method of speed control, leave the field coils connected directly to the power supply, and for the armature voltage control method, the field must be energized from a separate source.

in fact, the use of a field rheostat is the most popular method of controlling the motor speed because it is inexpensive, does not reduce efficiency, and has little effect on speed regulation (in percent).

The speed at which the motor runs with full load and rated voltage applied and with the field rheostat resistance set to zero is called the base speed of the motor. Adding resistance to the field circuit changes the motor torque curve, as shown in Figure 9-8, and increases the motor speed.[11]

At speeds below base speed, the shunt motor is like the permanent magnet type in that it can develop the same full-load torque at any speed. But when a field rheostat is used to increase the motor speed, the full-

[11]The other two methods of speed control reduce the speed below base speed.

load torque varies inversely with the speed. (The motor has a constant full-load power characteristic, as shown in Figure 9-8.)

Compound Motors

In addition to the shunt field coils, the compound motor has a set of series field coils connected in series with the armature and acting on the same poles as the shunt field. The shunt field is magnetically predominant (in fact, one can usually bypass the series fields of a compound motor, making it a shunt motor if desired), and so the same methods of speed control apply.

Cumulative compounding (connecting the series field so that it aids the shunt field) raises the starting torque obtainable with a given armature current (250 or 275% using

FIGURE 9-8 Torque characteristics of shunt-wound motors

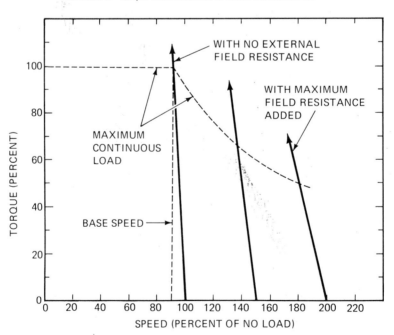

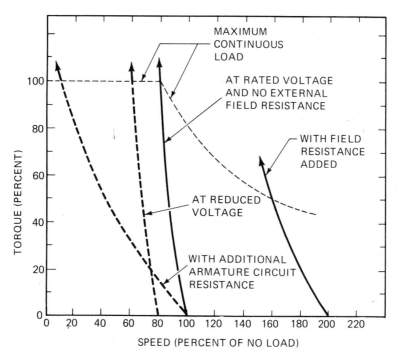

FIGURE 9-9 Torque characteristics of compound motors

200% current), but it increases the speed regulation to the range of 15 to 30%. Typical torque curves and the effects of each of the three possible methods of obtaining speed control are shown in Figure 9-9.

Differential compounding (connecting the series field so that it opposes the shunt field) improves the speed regulation at the expense of poor starting torque and/or unstable operation under some conditions, and is rarely done by industry. A packaged drive system would probably be used instead.

Series-Wound Motors

These motors have only series field coils. They have high starting torque (over 300% using 200% current) but very poor speed

regulation. At no load, a large series motor can speed up to the point where centrifugal force damages the armature (broken banding wires, leads pulled out of the commutator, or a commutator bar thrown out). For this reason the motor should be direct coupled or gear connected directly to its load.

The speed of these motors can be reduced by adding resistance in series or by reducing the applied voltage. As with the preceding types of motors, voltage control is efficient but expensive, and resistance control is cheap but inefficient. Typical torque curves and the effects of these speed controllers are shown in Figure 9-10. A few of these motors have taps in the series field coils that provide speed control by changing the external connections to the motor.

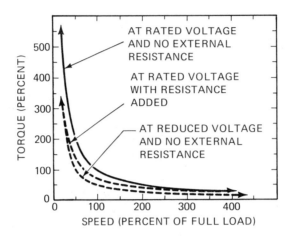

FIGURE 9-10   Torque characteristics of series-wound motors

THREE-PHASE SQUIRREL-CAGE MOTORS

Construction and Theory of Operation

These motors have two principal parts, the stator (so named because it is stationary) and the rotor (rotating part), which is mounted on a shaft supported on bearings.

When ac line power is applied to the stator windings, a revolving magnetic field is set up in the space occupied by the rotor. Depending upon how the stator windings are arranged, this magnetic field can have any even number of magnetic poles (adjacent poles being of opposite polarity) all evenly spaced, of equal and constant strength, and traveling around the stator bore at a constant speed. The speed at which the stator's magnetic field rotates is known as the synchronous speed of the motor. Synchronous speed is fixed by the frequency of the supply system and the number of magnetic poles created within the motor, and can be found by the Equation 4–1.

Synchronous speeds for 60-Hz squirrel-cage motors up to 14 poles are given in Table 9-4. It is rare to encounter motors with more than 14 poles; in fact 4- and 6-pole motors are by far the most common, while 2- and 8-pole motors are only used occasionally.

The revolving magnetic field in the motor does two things:

■ It generates a counter emf in the stator winding that is the main current-limiting factor there.
■ It induces currents in the squirrel-cage winding on the rotor.

TABLE 9-4   SYNCHRONOUS AND FULL-LOAD SPEEDS FOR 60-Hz INDUCTION MOTORS

Number of Poles	Synchronous Speed (rev/s)	Full-Load Speed (rev/s)
2	60	59
4	30	29
6	20	19
8	15	14.4
10	12	11.5
12	10	9.5
14	8.57	8.25

The rotor consists of a silicon steel laminated core keyed to a shaft. The rotor laminations are punched and assembled so that the rotor core has a number of closed or partially closed slots more or less parallel to the shaft and uniformly spaced around the outside periphery.

Copper or bronze bars are placed in the rotor slots and welded, brazed, or soldered to short circuiting end rings at either end of the rotor core. The bars and end rings form what is known as a squirrel-cage winding. A fan may be mounted on the shaft to circulate air for cooling the motor. On motors up to 500 kW or so, the squirrel cage and fan may be a one-piece aluminum casting.

When the revolving magnetic field created by the stator cuts the squirrel-cage bars, currents flow through the bars, which causes the following:

- Torque is produced, tending to pull the rotor along with the magnetic field. It should be emphasized that, if the rotor were moving at synchronous speed, there would be no cutting action, no rotor current, and therefore no torque. The difference between synchronous speed and rotor speed is referred to as slip. At no load the slip may be only 0.015 rev/s, but at full load, 1.0-rev/s slip is common. Unless the load is pushing the motor, it will have to run below synchronous speed. Typical full-load speeds of induction motors are given in Table 9-4.
- An increase of stator current: when rotor currents flow, they *tend* to modify the revolving field. However, when the revolving magnetic field changes, the counter emf changes and allows an increase of stator current. The increase of stator current practically offsets the mmf of the rotor currents, and so the magnetic field remains almost unchanged.

Characteristics of Induction Motors

Important Characteristics of Induction Motors

TORQUE–SPEED CHARACTERISTIC

- The starting torque developed is of interest because it indicates how much load the motor can start.
- The breakdown torque (maximum torque the motor will develop without stalling out or suddenly dropping in speed) gives a comparative indication of how well the motor will carry severe momentary overloads and/or operate during a severe momentary voltage dip.
- The full-load slip (the difference between synchronous and full-load speeds) indicates the amount of speed regulation the motor will have and also gives a comparative indication of how suitable the motor is for cyclic load applications.

CURRENT–SPEED CHARACTERISTIC   The shape of these curves is not the same for all motors, but the main point of interest is the current at zero speed, which is known as the starting current. This gives a comparative indication of how much disturbance will be created in the supply system when the motor is started. The current required by some motors remains near this locked rotor value until they are almost up to full speed.

Characteristics of Single-Speed Squirrel-Cage Motors

The NEMA (and EEMAC) have adopted a number of motor designs to meet users' requirements. The most common types are designs B, C, and D. The typical torque-speed and current–speed characteristics of these motors are shown in Figure 9-11(a)

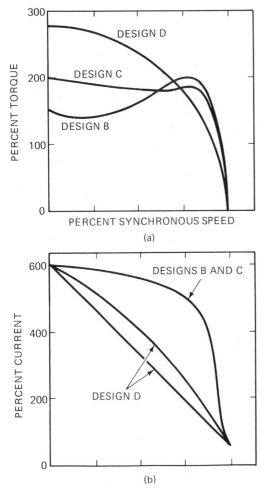

FIGURE 9-11   Torque and current characteristics of squirrel-cage motors

and (b), respectively. Substantially the same information is contained in Table 9-5. The reader should notice that the percentage values given refer to full-load current or full-load torque and are typical values for a 20-kW 30-rev/s motor. Smaller motors usually require more starting current (in percent) and may develop far more torque. Larger motors and/or low-speed types usually de-

velop slightly less torque and draw slightly lower starting currents.

The difference in the characteristics of these motors exists because of differences in rotor design. Specifically, design D motors have a high resistance squirrel cage (compared to design B), and design C motors usually have a double squirrel cage. These motors are often referred to by descriptive names. Design B are known as normal starting torque, low starting current motors. Design C are called high starting torque, low starting current motors, and design D are referred to as high starting torque, high slip motors. Design F is a low starting torque, low starting current motor.

Design B motors have the highest efficiency and lowest cost. Design C has slightly lower efficiency, design D efficiency is substantially lower, and both of these cost slightly more. So we use design B wherever possible and designs C, D, or F only where necessary.

Characteristics of Multispeed Squirrel-Cage Motors

Two-speed operation of squirrel-cage induction motors can be accomplished by

TABLE 9-5   **TORQUE AND CURRENT CHARACTERISTICS OF SQUIRREL-CAGE MOTORS**

	NEMA Designs		
Characteristics	*B*	*C*	*D*
Starting torque, %	150	200	275
Breakdown torque, %	200	190	–[a]
Slip (full load), %	2–4	2–4	5–8 or 8–13[b]
Starting current, %	600	600	600

[a] These motors do not have a breakdown torque as such; that is, they do not suddenly stall when overloaded. Maximum torque occurs at a standstill, and the torque curve has negative slope throughout the range from zero to synchronous speed.

[b] Motors with the higher range of slip have higher rotor resistance.

means of a single pole-changing winding (either conventional or pole amplitude modulation [PAM] design) or by the use of two separate stator windings connected for different numbers of poles. Three-speed motors are made with one conventional and one pole-changing winding, and four-speed motors have two pole-changing windings. The usual type of pole-changing winding always has a 2 to 1 ratio of speeds, but PAM motors and two-winding designs may have practically any ratio of speeds. All multispeed motors are available in three types:

■ *Constant power:* these motors have the same maximum permissible continuous power output at each full-load speed, and so the full-load torque developed is inversely proportional to the speed.
■ *Constant torque:* these motors have the same full-load torque at each full-load speed, so the power is directly proportional to the speed.
■ *Variable torque:* the full-load torque is directly proportional to each full load, and so the power is proportional to the square of the speed.

Some typical torque–speed curves for two-speed motors are shown in Figure 9-12. As with single-speed motors, the shape of the torque–speed curve depends on rotor design. Note that the general shape of the curve is the same on both speeds, and the starting and breakdown torques tend to be proportional to the full-load torque at each speed.

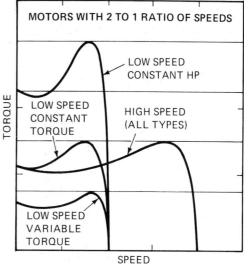

FIGURE 9-12   Torque curves for multispeed motors

WOUND-ROTOR MOTORS

Construction and Theory of Operation

The stator windings of these motors are the same as those of the squirrel-cage types.

But instead of a squirrel cage, the rotor has an insulated three-phase winding placed in its slots and connected into a wye, with its three terminals connected to a set of slip rings mounted on the shaft.

The supply lines are connected to the stator winding, and it therefore sets up a rotating magnetic field in the space occupied by the rotor. This induces voltages in the rotor winding, and if there is any closed path, current will flow in the rotor winding and torque will be developed. If the brushes that contact the slip rings are shorted together, the motor behaves like a low-resistance squirrel-cage type. But if resistance is connected between those brushes, the motor will perform like a high-resistance squirrel-cage design. As a result, slip-ring motors are known for three characteristics: high starting torque, low starting current, and the possibility of speed control.

Characteristics of Wound-Rotor Motors

With no external resistance in the rotor circuit (i.e., slip rings shorted together), a wound-rotor motor develops about 90% starting torque with about 500% starting current, and develops 250% breakdown torque at about 80% of synchronous speed. Adding resistance to the rotor circuit reduces the starting current and reduces the speed at which breakdown torque occurs. Suitable rotor circuit resistance will enable the motor to develop 250% starting torque using 350% starting current, more torque with less current than squirrel-cage motors require. If the rotor resistance is appropriately reduced during the acceleration period, the motor can develop 250% torque (using only 350% current) throughout the lower 80% of its speed range.

Additional rotor resistance reduces starting torque below the maximum, but further reduces starting current. Typical torque–speed curves and current–speed curves of wound-rotor motors with different values of

rotor resistance are shown in Figure 9-13(a) and (b). They are suitable for loads that have high inertia and/or require high starting torque and can operate over a range of speeds; but because of their cost, slip-ring motors are used only when a squirrel-cage motor cannot do the job.

Where wound-rotor motors are used with

FIGURE 9-13   Torque and current characteristics of wound-rotor motors

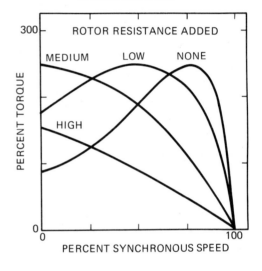

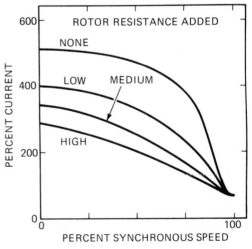

speed control, the maximum permissible continuous torque output (which the motor can develop without overheating) tends to be constant, and the maximum permissible continuous power output is reduced in direct proportion to the speed.

SINGLE-PHASE MOTORS

Although electrical power is generated and transmitted as three phase, most residential consumers and many light commercial users have only single-phase power available in the buildings they occupy. Regardless of where they are used, portable devices are generally single phase. For these reasons there are millions of single-phase motors in use. But in sizes larger than about 1 kW, they tend to be about 50% bigger and heavier than three-phase motors and correspondingly more expensive. Single-phase motors are also less efficient, and most are slightly more complicated mechanically so that three-phase motors tend to be more reliable. As a result, single-phase motors larger than 3 kW are quite rare, and sizes larger than 10 kW are practically nonexistent.[12]

Even in cases where a single-phase motor is available, a phase converter (or an appropriate service installation) and a three-phase motor might be more economical.

All the motors discussed in this section have pole-wound stators that are energized from the line and therefore have the same synchronous speeds as three-phase motors, and they all[13] have squirrel-cage rotors and practically the same theory of operation as

three-phase machines. Very few single-phase motors have more than eight poles. For convenience, typical synchronous and full-load speeds of single-phase motors are shown in Table 9-6. For practical purposes, the synchronous and no-load speeds will be the same.

Split-Phase Motors

These motors have two stator windings, each wound for the same number of poles and physically displaced from each other by one half of a pole width (90 electrical degrees). The starting winding is usually found near the top of the slots in the stator core. It is usually of small wire (to give it appreciable resistance), cotton covered (for improved ability to withstand overheating), and with relatively few turns of wire (to minimize its inductive reactance.) It is connected to the line through a centrifugally operated starting switch that opens up when the motor reaches about 75% of full speed. The starting winding is intended for only

[12]For applications involving large motors, use a three-phase motor and then weigh the economics of a single- to three-phase converter and wiring against the cost of installing the necessary service and wiring and obtaining the three-phase power from the utility company.

[13]The rotor of a repulsion start motor bears no physical resemblance to a squirrel cage, but in its normal running condition (short circuited) it is electrically the same.

TABLE 9-6   SYNCHRONOUS AND FULL-LOAD SPEEDS OF SINGLE-PHASE MOTORS

	Full-Load Speeds (rev/s)		
Poles	Shaded-Pole and Permanent-Split Capacitor Types	Other Types	Synchronous Speeds (rev/s)
2	—	57.5	60
4	27.5	28.75	30
6	17.5	19	20
8	—	14.2	15

momentary use and will quickly burn out if it is left connected to the line.

The running winding has heavier wire and more turns and is therefore bulkier than the starting winding; it has less resistance but more inductive reactance. As long as the motor is not overloaded, this winding can stand being energized continuously.

During the starting period, both stator windings are energized, but the current in the main winding lags the voltage by 60° or more, while the starting-winding current lags by only about 30°. Because of the time difference of the currents and the position difference of the windings, a rotating magnetic field is set up in the space occupied by the armature, and torque is developed thereon. When the motor has accelerated sufficiently, the starting switch deenergizes the starting winding, and the motor continues to run with just the main winding energized.

The most important features and limitations of split-phase motors are as follows:

■ Range of available sizes: about 50 to 400 W.
■ Torque characteristics: about 125% starting torque can be expected.
■ Speed regulation: usually less than 5%.
■ Speed control: not practical.
■ Current: starting current about six times the full-load current is usual.
■ Reversing: most motors can be connected to rotate in either direction. However, it is not possible to plug-stop these motors. On reversing applications, the motor must be allowed to coast to a stop, and then be energized for reverse rotation.
■ Inertia accelerating capability: not very good, because the starting winding overheats too quickly.
■ Reliability: not as good as three-phase motors, mainly because of the centrifugal mechanism and starting switch.

Capacitor Start Motors

These motors are similar to split-phase motors, but their starting windings have larger wire, and they have a large electrolytic capacitor wired in series with the starting winding. The larger wire in the starting winding reduces heating during acceleration, and the capacitor causes a larger phase difference between the starting and running winding currents so that the starting torque is improved. But running performance is no better than with a split-phase type.

■ Torque characteristics: the starting torque is usually 250% or more. The breakdown torque is usually somewhat lower, 150 to 200% being fairly common.
■ Current: full-load currents are about the same as for a split-phase motor, but starting currents are usually less, 500% being typical.
■ Reversing: ordinary capacitor start motors are limited in the same way as split-phase motors. But capacitor start motors built with a starting relay (instead of a centrifugal switch) can plug-stop and reverse.
■ Inertia accelerating capability: better than a split-phase motor but not as good as a three-phase type.

Two-Value Capacitor Motors

These are similar to capacitor start motors except that they have a small (capacity in microfarads) oil-filled capacitor connected in parallel with the large electrolytic type. When the motor starts, the starting switch disconnects the electrolytic capacitor, leaving the oil-filled capacitor and starting winding in series across the line. These motors are available in sizes from 1 to 5 kW, and performance is not greatly different from a capacitor start motor, but the power factor is improved, the motor runs quietly and the full-load line current is noticeably reduced. This last advantage makes it possible to run larger motors from a given size of branch circuit, which is particularly helpful in the case of higher-power residential or portable applications that must operate from 15-A branch circuits. Like the capacitor start, this motor can be plug-stopped, and reversed rotation is obtained by making the appropriate electrical connections.

Permanent-Split Capacitor Motors

These are similar to two-value capacitor motors except that they have no starting switch and electrolytic capacitor. The oil-filled capacitor is sized for optimum running performance. These motors have higher full-load slip and poorer speed regulation than split-phase motors; plug-stopping and/or reversed rotation can be obtained by making appropriate electrical connections. In addition to their quiet operation, these motors are suitable for speed control with fan loads. Reduced speed is obtained by reducing the applied voltage using an autotransformer, a series impedance (choke coil), or a triac.[14]

These motors are most commonly found in 150- to 250-W ratings. The absence of the starting switch makes them highly reliable, comparable to three-phase types.

Shaded-Pole Motors

A shaded-pole motor has only one insulated stator winding, which is wound on physically identifiable pole pieces or pole cores. However, it has uninsulated, short-circuited cop-

---

[14]If the load has constant-torque characteristics, do not attempt any type of speed control because it is almost certain to burn out the motor at reduced speeds.

---

per loops (known as shading coils or rings) enclosing an axial strip along one side of each pole. These shaded strips are about one quarter of the width of the pole. The flow of current in the insulated stator winding sets up an alternating magnetic field in the pole cores, and this induces currents in the shading rings. The combined effect of the main curent and the current in the shading coils is to set up a revolving magnetic field.

The important features and limitations of shaded-pole motors are as follows:

■ Available sizes: from about 10 to 200 W.
■ Torque characteristics: poor starting torque (about 70%).
■ Speed regulation: about 10%, because these motors usually have that much full-load slip.
■ Speed control: can be obtained with fan loads by reducing the applied voltage.
■ Current: rather high full-load currents for their size.
■ Reversing: unless specially designed, rotation can be changed only by disassembling the motor and reassembling it with the stator frame reversed relative to the shaft extension.
■ Inertia accelerating capability: not very good, mainly because the low starting torque prolongs the acceleration period.
■ Reliability: good, comparable to a three-phase motor.

Repulsion Start Motors

These have only one stator winding similar to the running winding of a split-phase motor. The armature has a commutator and an insulated winding connected thereto. However, the brushes that contact the commutator have no connection to the line, but are simply short circuited together. The armature will also have a centrifugally operated device that short circuits all the commutator bars when the motor comes up to speed, and sometimes there is a mechanism that lifts the brushes at the same time.

Repulsion start induction (RSI) motors develop tremendous starting torque (500% is not unusual) and will accelerate high-inertia loads without distress. Starting currents are about 600%, and speed regulation is good (less than 5%). But speed control is not practical, and since reversing the rota-tion is accomplished by adjusting the brush position, they are not suitable for applications involving frequent reversals. They are also somewhat more complicated (mechanically) than other single-phase motors and so may require more maintenance.

The RSI motor has been manufactured in sizes from 200 W to 7.5 kW. But because of its inherently high manufacturing costs, some prominent motor manufacturers have completely discontinued the type.

SYNCHRONOUS MOTORS

Construction and Theory of Operation

The stator windings of a synchronous motor are the same as those of an induction motor and, when energized, create the same kind of revolving magnetic field. But the rotor has an equal number of magnetic poles established by a dc winding, and when running these rotor poles are synchronized; that is, they lock into step with the stator magnetic poles. The stator poles travel at an absolutely constant speed. At no load the rotor poles line up almost exactly with the stator magnetic poles. When load is applied to the motor, the rotor poles will be pulled back behind the stator poles (this angle of lag is known as the torque angle); but as soon as the load is removed, the rotor poles will "catch up" to the stator poles again. We therefore say that the synchronous motor runs at a constant average speed.

The torque developed by a synchronous motor is roughly proportional to the torque angle. If the load is suddenly applied, the torque angle will increase; but because of inertia, the rotor will overshoot the necessary angle, then recover and undershoot, and so on, several times before settling down to a constant speed at the new torque angle. This momentary oscillation about the mean torque angle is known as hunting, and it is a normal occurrence when synchronizing (start up) or at any change of load. Usually this is not a problem, but with a pulsating load like a reciprocating air compressor, the load torque pulsations are sometimes close enough to the natural hunting frequency that continuous and severe hunting results. In some cases, the rotor poles "lose contact" with the stator poles, and the motor "slips," "pulls or falls out of step," or "loses synchronism."

When a synchronous motor is slipping, the rotor poles are alternately ahead and behind the stator poles having opposite magnetic polarity. The net torque developed is therefore zero; so once the motor slips it inherently stalls out completely and has no starting torque.

For starting purposes, most synchronous motors have a squirrel-cage winding on the rotor. When the stator is energized, they start up as induction motors, and then when direct current is applied to the field coils (the insulated rotor winding), the rotor poles lock into step with the stator poles. When the motor is synchronized, the squirrel cage cannot contribute driving torque, but it does help to squelch hunting. For this reason it is known as a damper winding or amortisseur winding. Large high-speed synchronous motors have cylindrical rotors, and the dc winding is placed in slots on the peripheral surface. But lower-speed machines have distinct pole cores wedged or bolted to a steel spider with a magnetizing coil around each core. When these low-speed machines are energized and approach synchronous speed, it is possible for the projecting (salient) pole cores to lock into step with the stator magnetic poles even before any dc power is applied to their windings. Torque produced in this manner is known as reluctance torque. This unexpected ability to synchronize with light loads has sometimes been a problem with synchronous motor drives.

One of the most important features of a synchronous motor is the way in which the dc current in the rotor winding influences the power factor of the stator circuit. With little or no dc current in the rotor winding, the stator winding draws a lagging current, the reactive component of which sets up the magnetic field in the motor. As the direct current in the rotor winding is increased, the reactive component of stator current decreases to zero and then reverses (i.e., it supplies lagging reactive power to the system) so that the motor is operating a leading power factor.

The dc power for the rotor of these motors has traditionally been obtained from a small dc generator (exciter) either direct coupled or belted to the synchronous motor shaft, and the connections to the rotor winding were made through two slip rings and brushes. But the advent of reliable solid-state diodes has made it possible to eliminate the slip ring and brush arrangement. On a brushless synchronous motor, the exciter is really a revolving armature stationary field type of three-phase alternator mounted on one end of the main motor shaft. The alternator output is rectified by a three-phase full-wave diode bridge mounted on the shaft and fed directly to the rotor winding of the synchronous motor. These brushless machines require less maintenance than ordinary types, and because all sparking is eliminated, they are particularly advantageous for use in certain types of hazardous locations.

Because they are more complex and require more elaborate control equipment than squirrel-cage machines, synchronous motors are rarely encountered in sizes below 200 kW. But in large sizes their cost becomes more competitive with squirrel-cage types. The power-factor advantage of synchronous motors has already been mentioned, but low-speed induction motors suffer from poor efficiency while synchronous motors do not, at least not to the same degree. Therefore, synchronous motors are most commonly found in low-speed designs of 500 kW or more.

Operating Characteristics of Synchronous Motors

Synchronous motors are not an "off the shelf" item but instead are more or less custom manufactured according to the user's requirements. If the constant average speed and/or power-factor correction of a synchronous motor make it attractive for a given application, the inertia and the torque characteristics of the load must be determined and made available to the motor supplier so that the required torque characteristics can be built into the motor.

PACKAGED DRIVE ASSEMBLIES

In the majority of motor applications, one can purchase an appropriate motor from one manufacturer, a suitable starter or controller from another, have a third party do the mechanical installation work, and use another contractor to do the electrical hook up. But where precise wide-range speed control and/or speed regulation of less than 1% is needed, standard motors and controllers will not do the job. For these applications one must use an automatic control system that will continuously sense the speed of the load and readjust the power input so as to compensate for variations of load torque, thereby holding the speed within an acceptable range of the desired value.

The problem is sufficiently complex that there is almost no hope of buying the necessary components from various suppliers and wiring them up into a successful system. But various manufacturers can supply complete systems of their own design to do the job. These packaged drive systems have all the components needed to convey the power from the ac line connection terminals right down to the input shaft of the driven machine, including the control system. They usually include a fair number of electronic devices like transistors, UJT's, triacs, SCR's, and so on. Larger ones will require a three-phase supply, but smaller units may be designed for single-phase operation.

Solid-state electronic circuits are highly reliable, and so the overall maintenance requirements are not much greater than those of the drive motor involved. But packaged drives may need careful adjustment on the initial installation, and when a failure occurs, the average electrician has neither the knowledge nor the equipment needed to find the trouble and correct it promptly. So it is important that the manufacturer's service representative be quickly available when needed.

Squirrel-Cage Motor and Eddy-Current Clutch

The motor in these systems is a more or less standard squirrel-cage type that is directly energized from the line and runs continuously. But the motor is coupled to the load by means of an electrically operated slip clutch arrangement that has no wearing parts except bearings. A speed-sensing device on the output shaft feeds back to the control system, which compares this signal to the desired value and readjusts the power supplied to the clutch coil, thereby increasing or decreasing the slip. To obtain fast slow-down action and to handle overhauling loads, these drives may include an automatically controlled eddy-current brake as well. Sizes from $\frac{1}{4}$ to 75 kW are available.

The main features of these drives are as follows:

■ *Excellent speed regulation:* in most cases the speed regulation can be adjusted to less than 1%. In some cases, one can obtain zero or even negative speed regulation, but in other cases the drive may suddenly begin to hunt if the speed regulation is set too low.

■ *High starting and accelerating torque is available:* since the motor starts before the clutch is energized, the maximum torque delivered to the load is limited to either the maximum torque that the clutch can transmit or else the breakdown torque of the motor, whichever is smaller. If the clutch could be applied suddenly, one could use the motor inertia to start the load, but most control systems are designed to prevent "shock loading" of the drive in this way.

■ *Poor efficiency:* the power consumed by the control coil of the clutch is not all that much, but the power loss due to clutch slippage is considerable, particularly at reduced speeds. The resultant heating of the clutch may limit the maximum permissible continuous torque loading at very low speeds to something less than the full-load torque of the motor; in some cases, the clutch may have to be water cooled.

Squirrel-Cage Motor and Variable-Frequency Power Supply

Speed control is obtained here because the speed at which an ac motor runs is directly proportional to the frequency of the power supplied to it. Both induction and synchronous motors have been used, but ordinary motors are designed for operation at only one frequency, so the motors are usually a special design. These systems have good speed regulation and high efficiency throughout the speed range. The variable frequency power is obtained in either of the following ways:

Cycloconverter

Cycloconverters change the line frequency ac power to low-frequency ac power by means of appropriately timed electronic switching operations. The output frequency is controlled by changing the rate of the switching operations and can be varied from zero (dc) to about 20 Hz. This system is best adapted to low-speed drives and has

been used in sizes up to several thousand kilowatts.

Rectifier and Inverter Assembly

The rectifier changes the line frequency alternating current to direct current and the inverter "chops up" the direct current to provide alternating current at the desired frequency. This system has no special frequency limitations, and some systems have operated at up to 800 Hz, giving a two-pole motor a synchronous speed of 800 rev/s.

When a squirrel-cage motor is operated at reduced frequency and voltage, starting and breakdown torques are reduced as shown in Figure 9-14. As a result, the system is inherently suitable for fan loads and must be somewhat oversized for continuous operation of constant-torque loads at reduced speed.

Squirrel-Cage Motor with a Variable-Voltage Power Supply

Reducing the applied voltage lowers the torque curve throughout the speed range of the motor, and so with a given load, the motor speed will decrease. But, unfortunately, the rotor copper losses are directly proportional to load torque and directly proportional to the motor slip. So if the motor size is appropriate for high-speed operation and this form of speed control is attempted with a constant-torque load, serious overheating of the squirrel cage will occur at low and medium speeds. Regardless of the relative size of the motor and load, it is impossible to get stable operation at speeds below that at which breakdown torque occurs (motor stalls out).

With fan loads the situation is different. Since the torque requirement of a fan is low, squirrel-cage overheating at low speeds is

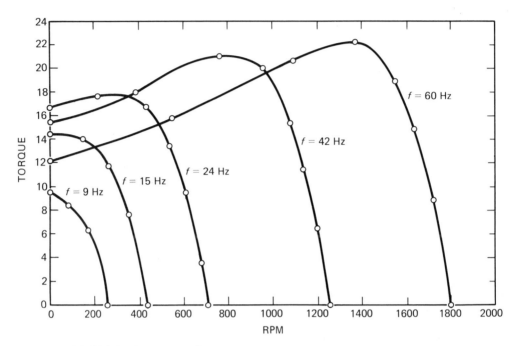

FIGURE 9-14 Torque curves for a squirrel-cage motor on adjustable frequency power supply

eliminated. And if the motor torque curve has negative slope throughout the speed range, overheating at medium speeds is prevented,[15] and stable operation at any speed will be assured.

Motors used with a variable-voltage power supply are designed for the application and have torque characteristics similar to NEMA design D. The reduced voltage is commonly obtained through delayed firing of SCR's

or triacs. With a suitable control system, this scheme can provide excellent speed regulation, wide-range speed control, and timed or current-limited accelerating, but the efficiency tends to be rather poor at reduced speeds.

DC Motor and Adjustable Power Supply

These systems generally use a shunt-wound dc motor with a fixed voltage applied to the field and an adjustable-voltage power supply for the armature. The earliest form of this drive used a standard dc motor fed from a conventional dc generator (M–G set) and is known as a Ward–Leonard system of speed control. Later versions used an amplidyne or rototrol (these are special types of generators that have quick response and are readily

---

[15]One might be tempted to use a three-phase variac to control the speed of a motor-driven fan. However, if this is attempted with a NEMA design B (or C) motor, squirrel-cage overheating can occur at intermediate speeds, so the method is not recommended. But if connected to a fixed low-voltage source, these motors can operate a fan at low speeds without harm. We sometimes use this method to get high speed–low speed operation of a fan where on–off operation is not practical (due to icing problems for example).

adapted to automatic control) or a set of control system, speed regulation is excellent (comparable to the squirrel-cage motor and eddy-current clutch), a wide range of speed control can be obtained, and other features such as timed acceleration or torque-limited operation are also obtainable.

adapted to automatic control) or a set of thyratrons (a tube type of rectifier whose conducting time per cycle can be adjusted), but the newest types use silicon-controlled rectifiers. The available range of power ratings extends from $\frac{1}{2}$ to 200 kW.

It is obvious that the maximum permissible starting and continuous output torque of these drives cannot exceed the inherent capabilities of the incorporated motor. Since the speed control is obtained by the armature voltage method,[16] efficiency tends to be high throughout the speed range. Because of the control system, speed regulation is excellent (comparable to the squirrel-cage motor and eddy-current clutch), a wide range of speed control can be obtained, and other features such as timed acceleration or torque-limited operation are also obtainable.

In general, the use of a rectifier as a power supply precludes regenerative braking of a dc motor. If such slowdown is required, or if the system is to be used with an overhauling load, an automatically controlled dynamic braking circuit will have to be included in the package.

SUMMARY

Most interior wiring systems operate with alternating current and may be either single or three phase. To ensure compatibility among the various components, they operate at one of only a few preferred standard voltages.

If properly interpreted, the nameplate of a motor yields considerable information—not only electrical data but also such data as speed, temperature rise, and power factor.

Electric motors are available in either single- or three-phase ac types and dc motors, also. In general, three-phase motors are used in larger sizes and/or industrial applications, while single-phase motors are used for small and/or residential applications. For some purposes, dc motors or packaged drive systems will be required, but these are only found in an industrial setting, driving loads for which ac motors are unsuitable.

REVIEW QUESTIONS

9-1. In terms of voltages, what is the difference between a single-phase two-wire and a single-phase three-wire circuit?

9-2. What is the difference between a single-phase three-wire and a three-phase three-wire circuit?

9-3. What is the difference between a four-wire wye and a four-wire delta system?

9-4. What is the advantage of a one-line diagram?

9-5. What is meant by the terms "line drop," "voltage flicker," and "voltage regulation"?

9-6. What magnitude of voltage flicker is generally objectionable?

9-7. How much voltage variation (in percent) is tolerable at the load under normal conditions?

[16]Some of these drives also include field flux control.

9-8. Briefly explain how electrical fires get started.

9-9. What purposes are served by the boxes used to enclose electrical equipment?

9-10. What types of electrical enclosures are available?

9-11. In what three forms are electrical circuit conductors manufactured?

9-12. Other than circuit conductors and the motor, what are the four essential components of a motor branch circuit?

9-13. Why is it important to have a disconnect switch within sight of a motor?

9-14. What type of a single-phase motor is not called a squirrel-cage motor?

9-15. What is the significance of the following data when found on a motor nameplate?
(a) voltage
(b) current
(c) kW
(d) number of phases
(e) frequency
(f) speed
(g) designation of terminals
(h) evidence of approval
(i) enclosure
(j) insulation class
(k) temperature rise
(l) service factor
(m) frame size
(n) code letter
(o) power factor

9-16. What are the distinguishing features of open, dripproof, TEFC, TENV, and explosion-proof motors?

9-17. What can the user do to protect a dripproof motor from condensation when it is not operating?

9-18. There are four basic methods of connecting a motor to its load. What are they and what are the inherent advantages and restrictions of each?

9-19. What is meant by a resilient base and why is it used with single-phase motors but not three-phase types?

9-20. (a) For what mounting position are most motors intended?
(b) Can standard motors be mounted on a wall or ceiling; if so, what alterations might be desirable?
(c) Vertical shaft applications should be checked with the motor manufacturer. Why?

9-21. On what basis are dc motors classified and what types are commonly available?

9-22. What type of dc motor:
(a) Has the best speed regulation?
(b) Has the highest starting torque?
(c) Will run away under no-load conditions?

9-23. In what two ways can the speed of a permanent magnet motor be controlled, and what effect does each method have on the torque curve?

9-24. In what three ways can the speed of a shunt-wound motor be controlled, and how does each method affect the torque curve?

9-25. Speed control can be obtained on most series-wound motors in either of two ways. How does each method affect the torque curve?

9-26. Single-speed three-phase squirrel-cage motors are available in NEMA designs B, C, D, and F. How do the torque and starting-current characteristics of these motors compare?

9-27. Multispeed squirrel-cage motors are available in constant-torque, constant-power, and variable-torque designs. Compare their respective torque characteristics at each speed.

9-28. What are the three performance features of wound-rotor motors?

9-29. What must be done to change the shape of the torque curve of a wound-rotor motor?

9-30. What types of single-phase motor:
  (a) Have the highest starting torque?
  (b) Can be plug-stopped?
  (c) Can be used for speed-controlled applications?

  (d) Can be conveniently reversed by means of electrical switching?

9-31. What are the two advantages of synchronous motors?

9-32. What four kinds of packaged drive assemblies are available?

9-33. Which of the packaged drive systems have the best low-speed efficiency?

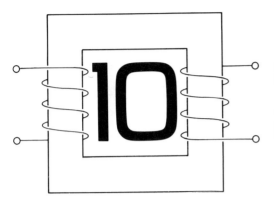

MOTOR STARTERS AND CONTROLLERS

For most applications we must have some convenient way to start and stop a motor, and sometimes there is a need to reverse the direction of rotation and/or control the speed. The switching devices used to perform these tasks are known as motor starters or motor controllers, and manually operated switches are commonly used. However, electrically operated switches known as relays or contactors are more suitable for switching large motors or for switching operations requiring more than one step, and they are absolutely essential if the motor is to be controlled automatically or by a computer. We will therefore begin with a brief description of relays and how they can be controlled. Since time-delay relays are very useful in the motor-control business, we will cover them as well.

With very few exceptions, all motors must be provided with overload protection. For that reason we will briefly describe the four common types of overload protective devices and outline a few advantages and disadvantages of each.

Because of their similarities and differences, we have grouped the starters and controllers according to the type of motor they control, and all common varieties have been included.

Emphasis has been placed on the power circuits and the required cycle of operation because these tend to be common to all manufacturers.

Some representative control circuits have been included, but a great many variations are in use and it is not possible to include them all here.

BASIC PRINCIPLES OF RELAY CONTROL

Relay Construction

In its simplest form, an electromechanical relay consists of an assembly of contacts (held in a position called "normal" either by gravity or by a spring) and an electromagnet capable of pulling the contact assembly away from the normal position. It is a magnetically operated switch; in fact, the terms "magnetic switch" and "magnetic contactor" are commonly applied to some heavy-duty types of relays.[1] The contact assembly usually has one, two, three, or four poles (although relays of up to 60 poles have been manufactured) and may include normally open and/or normally closed contacts on each pole. Although there are a few exceptions, most relays have only one actuating electromagnet coil, and that is the only type we will consider.

Relays have been in use for over 70 years and millions of them are manufactured and sold annually. As a result, many refinements of design have gradually taken place; but the operation of a modern solid-state relay, or a reed relay, or a vacuum contactor can be readily understood in terms of a simple electromechanical type, and the control circuits required are all basically the same.

Relays in Schematic and Wiring Diagrams

In a wiring diagram we try to show all components in their actual relative positions so that the diagram will resemble the real equipment as closely as possible. A small TPDT relay is shown in Figure 10-1, and it should be emphasized that the contacts are shown in their normal position.

If a drawing contains more than one relay, each relay is identified by means of letters or numbers or combinations thereof, such as M, $CR1$, $CR2$, $1A$, $2A$, and so on, and wherever possible these designations are selected and applied so as to remind the reader of the relay's actual function.

[1]If the contact assembly is rated at 10 A or less, we generally use the term "relay." If the contact assembly is rated at more than 10 A, the word "contactor" is preferred.

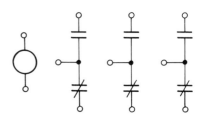

FIGURE 10-1   Three-pole double-throw (TPDT) relay

Wiring diagrams are useful for construction or installation purposes; but if it becomes necessary to explain how a circuit works, a schematic diagram is more convenient. In a schematic diagram, the parts are arranged so that the interconnections can be shown in the simplest form without regard to the actual physical arrangement of the parts. Contacts from different poles of a relay may be widely separated (on the drawing) from each other and/or from the coil that actuates them. But identifying letters or numbers (the same as those used on a wiring diagram) are shown beside each coil and each contact so that the operation of the circuit can be clearly understood. Both schematic and wiring diagrams are to be found in subsequent sections of this chapter.

Elementary Relay Control Circuit

Figure 10-2 shows a possible control circuit for a relay that has a 6-V dc operating coil. As long as the toggle switch is open, the relay contacts will remain as shown. But if the toggle switch is closed, the coil will pull the contacts closed, and the load will then be energized. When the coil is energized, we say that the relay picks up (meaning that it moves away from the normal position). When the toggle switch is opened, the coil will be deenergized and the relay will release (return to the normal position), and this in turn will deenergize the load.

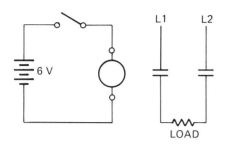

FIGURE 10-2   Elementary relay control circuit

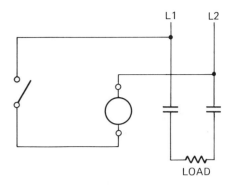

FIGURE 10-3   Relay control circuit energized from the power circuit

The relay contacts and the wiring that carries the load current are known as the power circuit. The toggle switch, the coil, and the wiring that carries only the coil current are collectively referred to as the control circuit or pilot circuit, and the arrangement in Figure 10-2 is commonly known as a two-wire control circuit.

A device that controls the operation of a relay is commonly known as a pilot device. In Figure 10-2, the toggle switch serves as a pilot device. There are many other devices, such as thermostats and pressure switches, that may be used in the control circuit. Because of the nature of their action (once positioned they stay put until some definite change causes them to move again), these are known as maintained contact pilot devices. Maintained contact pilot devices always result in a two-wire control circuit.

If the voltage rating of the relay coil is the same as that of the power circuit, the control circuit may be energized therefrom (see Figure 10-3). The operation of this circuit is identical with that of Figure 10-2.

Some Advantages of Relay Control

The fact that a single-pole pilot device will control a relay that has a large number of poles can be applied in two ways. A multi-pole relay will enable us to simultaneously switch all ungrounded conductors in a power circuit without resorting to multipole pilot devices. But it is also possible to have each relay pole controlling a separate circuit so that a single-pole pilot device can control a multitude of circuits. The relay in this case serves as a "circuit multiplication" or "fanout" device. Normally closed contacts on a relay can also be used to invert the action of a pilot device, as shown in Figure 10-4.

When the pilot device is closed, the relay contacts will be open, and when the pilot contacts are open, the relay contacts will close. If this circuit is applied to a thermostat whose contacts close on temperature fall, the relay contacts will close on temperature rise, and so the thermostat's action will have been effectively inverted.

The fact that the control circuit can be

FIGURE 10-4   Inverting the action of a pilot device

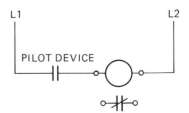

completely isolated from the power circuit of a relay (and their voltages therefore do not have to be the same) is also advantageous. It enables use to interface (interconnect or "tie together") systems of different voltages so that logic devices operating at 5 V can control a relay that switches 120 or 240 V, and it permits us to control motors operating at 240 V or higher, using only 120 V or less on the pilot circuit.

It should also be noted that with few exceptions the current capacity of the contacts on a relay is much higher than the current in the relay coil. The difference between the voltage and current ratings of the control and power circuits is so important that we will quote a few examples in Table 10-1 and then discuss the advantages that this difference presents.

Let us consider the size 5 contactor listed in Table 10-1. An equivalent manually operated switch would require considerable force to operate it (on the order of 200 newtons or so). From the standpoint of operator convenience, it is certainly preferable to use the contactor and control it with a manual switch that requires about

1 newton of actuating force. A closely related advantage arises from the fact that limit switches, thermostats, and similar switching devices cannot be manufactured with current capacities exceeding a few amperes. However, relays and contactors permit such small switching devices to control large loads and therefore make possible (and are essential to) automatic control.

Since it carries only a small current, the control circuit wiring can be of a small cross-sectional area and therefore of moderate cost per meter. Voltage drop and power loss will also be small, so that several hundred meters of control circuit wiring is practical. Power circuit wiring, however, is often of rather large cross section and is correspondingly more expensive. If a contactor is used, it can be located so that the power circuit wiring will be as simple and direct as possible, and the control wiring will run to the desired control location(s). But if the load is to be directly controlled by manually operated switches, then the power circuit wiring must be run to the control location(s), and this may increase the cost, particularly if it is desired to control the load from two or more locations. Line loss and line drop problems will also be increased by such an arrangement. So a contactor provides an economical way to obtain remote control (control at a distance), particularly from more than one location.

TABLE 10-1   VOLTAGE AND CURRENT RATINGS OF CONTROL AND POWER CIRCUITS

Device	Power Circuit		Control Circuit	
	Voltage (V ac)	Current (A)	Voltage (V ac)	Current (A)
Typical small relay	250	10	120	0.025
EEMAC size 5 contactor	600	270	240	0.41
Vacuum contactor	5000	360	120	1.0

Relay Control with Momentary Contact Push Buttons

Referring to Figures 10-2 and 10-3, if a normally open push button is substituted for the toggle switch, the resulting circuit is of little practical use, because the load will be energized only while the push button is held down. It would be more convenient if the relay would remain energized once it has picked up, provided that we can cause it to release when desired. The desired holding action can be obtained by connecting a pair of contacts from the relay in parallel with the push button, as shown in Figure 10-5. The relay contacts used for this purpose are known as holding contacts, maintaining contacts, or seal-in contacts.

It is obvious that the relay will remain in the released position until the button is pushed, and once the relay has picked up, it will remain so even if the button is released. But once it has picked up, there is no convenient way to make the relay release, and so the circuit shown in Figure 10-6 has been developed. This circuit is known as the conventional lock-up circuit or a three-wire control circuit. More complex push-button

FIGURE 10-5 Partially developed relay control circuit

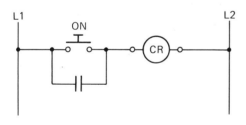

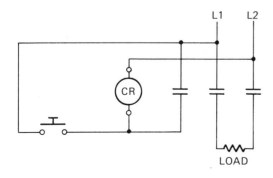

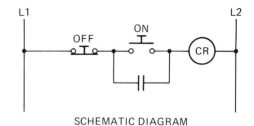

SCHEMATIC DIAGRAM

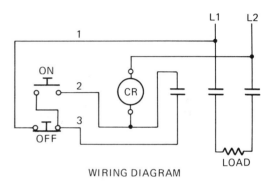

WIRING DIAGRAM

FIGURE 10-6 On-off push-button control of a relay

control circuits are used, but most of them are derived from this simple scheme.[2] Whenever momentary contact pilot devices are used, a three-wire control circuit usually results.

It can be readily seen that if both the *on* and *off* buttons are pushed simultaneously, the relay will release and/or remain so. We describe this action by saying that the off button overrides the *on* push button in this circuit (i.e., when both buttons are pushed, the relay obeys the *off* push button); or alternatively we say that the *off* push button will release and lock out the relay (i.e., with the *off* button held down, the relay will not pick up if the *on* button is pushed). Other

[2]The relay lock-down circuit and its variations are exceptions, but they are not commonly used.

circuit designs are possible, but for motor control this is the preferred arrangement.

Classification of Control Circuit Contacts

Let us reconsider the circuit in Figure 10-6. If we were to replace the *off* push button with several normally closed buttons connected in series, any push button in that group could still release and lock out the relay. To be a little more general, if we replace the *off* push button by any combination of switches, push buttons, or contacts on other relays, the fact remains that as long as the group constitute an open circuit the relay will still release and/or be locked out. We can therefore refer to that group of contacts, switches, and so on, as the turn-off set or turn-off group (or lock-out set) of contacts, and we will do so in subsequent portions of this section.

In similar fashion we can visualize the *on* push button as being replaced by a group of contacts called the turn-on set, and if this set constitute a closed circuit (assuming the turn-off set also form a closed circuit), the relay will pick up and hold.

The holding action obtained by using a pair of contacts on the relay to bypass the turn-on set is fundamentally different from the action of either the turn-on or turn-off set. But we could add other switches or contacts in series with the holding contacts of the relay and therefore make the holding action dependent upon continuity through the whole group. We call that group of contacts (including those on the relay) the holding set. These three groups of contacts[3] are shown schematically in Figure 10-7.

---

[3]The terminology here is patterned after D. Zissos, *Logic Design Algorithms*, Harwell series (New York: Oxford University Press, 1972).

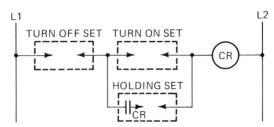

FIGURE 10-7 Classification of control circuit contacts

If the circuit is left in its primitive form, every contact in any relay control circuit can be readily classified as belonging to one of these three sets. To minimize the number of contacts required, we sometimes modify a control circuit into a form that blurs these classifications but does not eliminate them. The two-wire control circuit is a special case of a circuit that has no holding set, so there can be no distinction between the turn-on and turn-off sets; and in some circuits the turn-off set does not exist, but only the turn-on and holding sets.[4]

Interlocked Relays

If the status or position of one relay affects the operation of another relay, the two relays are said to be interlocked. If the interlocking is accomplished by means of a mechanical linkage, the relays are said to be mechanically interlocked; but if it is accomplished by connecting a set of contacts on one relay into the control circuit of the other, the relays are said to be electrically interlocked.

A mechanical interlock is commonly used to prevent two relays from being in the picked-up position at the same time, but that seems to be the limit of its practical applications. Electrical interlocking can be

---

[4]This is still a functional circuit since the relay will release if both the turn-on and holding sets are opened.

arranged to obtain the same effect, and it also offers many other possibilities.

Considering two relays *A* and *B*, if the status of relay *A* affects the operation of relay *B*, but not vice versa, we will so indicate by saying that the relays are "unilaterally interlocked"[5] with *B* dependent upon *A*.

Let us assume that the two relays *A* and *B* each have an on–off push-button control circuit. If it is desired to unilaterally interlock the relays so that *B* is dependent on *A*, 10 unique arrangements are possible using only one contact on relay *A*. The possible variations are as follows:

- The use of either a normally open or normally closed contact from *A* in the control circuit of *B*.
- The location of that contact as part of either the turn-on, turn-off, or holding sets of the control circuit for *B*.
- Connecting the contact from *A* either in series or parallel with the existing contact group in the control circuit of *B*.[6]

If the status of a relay *C* affects the status of some other relay *D*, and vice versa, we will so indicate by saying that the relays *C* and *D* are "bilaterally interlocked."[7] And if the status of *C* affects *D* the same way as *D* affects *C*, we designate this as a "symmetrical interlock" arrangement.[8]

---

[5]The meaning attached to the words in quotation marks is peculiar to this text. There does not appear to be any widely recognized terminology that conveys the intended meaning.

[6]Since $2 \times 3 \times 2 = 12$, one might have expected 12 possible arrangements. But since a contact in parallel with the existing holding circuit becomes a part of the turn-on set, two pairs of the 12 combinations are identical, and so there are only 10 *unique* arrangements.

[7]See fn. 5.

[8]A mechanical interlock between two relays is usually a symmetrical arrangement.

Using only one interlocking contact from each relay, two relays can be bilaterally interlocked in 55 different[9] ways, of which 10 will be symmetrical. However, not all these possible combinations result in functional circuits.

So far we have considered interlocking of relays using distinct contacts that serve no other purpose.[10] This is not an unusual practice by any means, but if the interlock contact is of the same type as some other contact on the same relay (e.g., both normally open), it is frequently possible to rearrange the circuit so that one contact serves two purposes and thus to reduce the total number of contacts required. And if the status of the relays is considered unimportant and only the energization (or otherwise) of the loads is significant, the performance of some interlocked relay control circuits can be effectively duplicated by interconnecting the power circuits of the two relays. It is convenient, however, to consider all such refinements as special cases.

Overlapped Contacts

If a relay has both normally open (NO) and normally closed (NC) contacts, the usual arrangement is such that, as the relay picks up, the NC contacts open first and then the NO contacts close.

However, for some control circuits it is necessary to have the NO contacts close before the NC contacts open.

---

[9]Because of the obvious triviality, we will not consider reversing the roles of the two relays, which would bring the total to 100.

[10]Heavy-duty contactors usually include one or more pairs of light-duty contacts known as auxiliary contacts to act as holding contacts and/or for interlocking purposes.

Two pairs of contacts (one NO and one NC) arranged to obtain this action are known as overlapped contacts.

On motor-control equipment, you may find a NC contact that does not open until the NO contacts have closed.

The NC contacts in such a case will be marked LB, which stands for "late break."

SPECIAL FEATURES OF COMMERCIAL MOTOR STARTERS AND CONTROLLERS

A motor starter is a switching device that is intended to start and accelerate a motor. A motor controller is a device that controls the power flow to a motor, which usually results in some form of speed control.

Some motor starters perform only a one-step (on–off) switching operation, but some starters (and most controllers) go (or can go) through a number of switching operations in a definite sequence and with time delays between the steps. If the starter or controller is manually operated, it is generally constructed so that it cannot be operated in the wrong sequence, but usually the operator must provide the time delays required for each step. If the starter is an automatic type, the switching operations are carried out by contactors (heavy-duty relays), and if it goes through a sequence of steps, the required time-delay devices are built in as part of the unit. We will consider these time delay devices in more detail next.

In addition, most starters and controllers include one or more devices known as overload relays that will protect the motor from overheating due to excessive mechanical load. We will examine these devices in detail, followed by consideration of the standard terminal markings for starters and controllers and then economizer circuits.

Time-Delay Techniques

An ordinary relay will pick up or release in perhaps $\frac{1}{4}$ second or less. The simplest time-delay devices are basically relays on which either the pickup or release motion (rarely if ever both actions) has been purposely slowed down. If the pickup action is slowed down, we call it an "on-delay" timer; if the release action has been slowed down, we call it an "off-delay" timer. Either of these timers may have both normally open and normally closed contacts. Some timers have certain contacts that are time delayed and other contacts that have no delay.

To prevent confusion, we normally show on a diagram what movement (if any) is delayed by placing the markings TC (timed-closing action) or TO (timed-opening action) beside each timer contact. If a particular contact is not delayed, these markings do not appear. We always show the timer contacts in the deenergized (normal) position, and as a result one can tell by noting the action of any timed pair of contacts whether the timer is an on-delay or an off-delay type. If timed closing contacts are shown open (or timed opening contacts are shown closed), it is an on-delay timer. If TC contacts are shown closed (or TO contacts are shown open), it is an off-delay timer. See Figure 10-8.

There are various ways of delaying the mechanical motion of the contacts in a timer. An oil-filled dashpot or an air-bellows arrangement will work effectively. Some manufacturers use an escapement mechanism (a

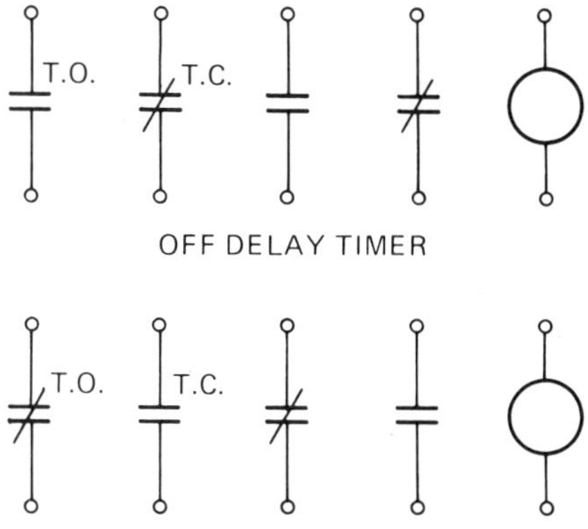

OFF DELAY TIMER

ON DELAY TIMER

FIGURE 10-8 On-delay and off-delay timers

ratchet and oscillating weight assembly closely related to the escapement and balance wheel mechanism of a spring-driven clock); others use a cam-operated switch and drive the cam with a small motor. The manufacturer can slow down the action of a dc delay by inserting a copper sleeve inside the electromagnet coil. This technique is particularly effective in obtaining slow release action; a delay of several seconds can be obtained.

Overload Protection

If there is too much mechanical load on a motor, it draws excessive current from the line and therefore overheats. An overload relay is a device intended to protect a motor from this condition. There are four types of overload protection schemes in use:

Thermal Overload Relays

These devices have a small heating element connected in series with the motor. If sufficient heat is produced by the heating element, it will open a pair of contacts, and so these devices sense the motor current rather than its temperature. Figure 10-9 shows a three-pole overload relay assembly where heat from any one of the heaters will open the contacts.

Some thermal overload relays use a bi-metallic strip to actuate the contacts. Once the contacts open, some of them will reclose (or reset) automatically, while others must be reset by manually pushing a button or lever. Others depend upon the melting of a eutectic alloy (solder film) to release the contacts (which are spring loaded to open), but this type has to be reset manually.

To match the overload relay with the motor it is supposed to protect, most manufacturers can supply a series of heater elements (or heaters as they are called) with current ratings of 0.5 A and up in steps of about 9%. It is therefore possible to obtain overload protection set within the range of 105 to 115% of the full-load current of any motor just by selecting the proper heater. If we select a heater one size larger, the overload protection will be set between 115 and 125% of the motor full-load current. Normally, the manufacturer will supply a table that shows the catalogue number of the heaters that should be used with motors of various current ratings. When selecting heaters from that table, follow the manufacturer's directions carefully.

Thermal overload relays are by far the most popular form of overload protection;

FIGURE 10-9 Three-pole overload relay

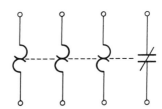

they will protect the motor against excessive current, but they have a few shortcomings. First, they inherently trip with less current if the ambient temperature is high and therefore are subject to false tripping. Ambient-compensated overload relays overcome this problem. Second, since they respond only to motor current, thermal overload relays cannot protect a motor against high ambient temperature or inadequate ventilation. Third, the time–current characteristics of these relays are usually not adjustable. To get faster or slower tripping action, we have to choose a different model of overload relay that frequently takes a different series of heaters. The result is a proliferation of different heaters that must be manufactured and stocked in order to meet the customer's requirements.

Magnetic Overload Relays

These overload relays are actuated by an electromagnet coil that attracts an iron core. The coil is wired in series with the motor, and the movement of the core is delayed by an oil-filled dashpot or by an air-bellows arrangement. Like the thermal type, these relays respond to motor current.

The pickup current and/or the time delay of a magnetic overload relay is usually adjustable in the field so that only a limited number of different sizes and types is required.

Many magnetic overloads have a rather rapid reset action, often less than 5 s. If such an overload relay is slow enough acting to permit the motor to start (and it obviously must be so), it will permit any number of starts in quick succession. For this reason some magnetic overload relays cannot protect a motor against burnout due to repetitive starting.

Electronic Overload Relays

These sense the voltage applied to the motor, as well as the current flow through it, and electronically simulate both the iron loss and copper loss within the motor. The pickup current and time-delay characteristics are both usually adjustable, and so these relays can provide protection that is very close to (but not exceeding) the limitations of the motor. Most of these overload devices are inherently immune to ambient temperature effects. They can be adjusted to provide either manual or automatic reset and can provide other special features, such as instantaneous protection against single phasing of a three-phase motor.

Thermistor Types of Overload Protection

A thermistor is a special type of resistor whose resistance increases substantially as the temperature increases. For motor protection, we need thermistors whose resistance increases sharply at a temperature just below the maximum temperature rating of the insulation used in the motor. If we embed such thermistors in the motor winding and monitor their resistances electronically, a sudden increase of resistance means that the motor is getting too hot, and the electronic circuitry can be arranged to shut off the power supply.

Thermistor schemes can protect a motor against high ambient temperatures or inadequate ventilation and can be adjusted for either manual or automatic reset. They are best suited for three-phase motors, but thermistors must be embedded in the stator windings when the latter are installed. Because of the difficulty of making connections to thermistors located on the armature, they have rarely if ever been applied to dc machines.

Terminal Markings

The line terminals are generally marked $L1$, $L2$, and so on. The load terminals are generally marked to correspond with the terminal markings on the motor it is intended to control. On a dc motor starter, you will find that $A1$ and $A2$ indicate the armature connections, $F1$ and $F2$ indicate the shunt field connections, while $S1$ and $S2$ indicate the series field connections. On three-phase motor starters, $T1$, $T2$, and $T3$ will appear, and possibly others if the motor requires more than three line connections.

All automatic starters are designed for three-wire control circuits, and the control terminals are generally marked with plain numbers (no letters). The stop-button connection is number 1, while number 2 provides the holding circuit or path and therefore must connect to both the start and stop buttons. Terminal 3 connects to the start button only as shown in Figure 10-6.

Economizer Circuits

To develop tractive force, the magnetic circuit for the coil of a relay must necessarily have an air gap that closes as the relay picks up. The closing of the air gap permits the magnet coil to produce more flux with the same current, and so if the magnetic pull

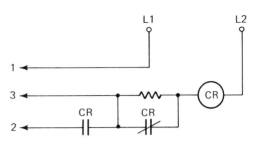

FIGURE 10-10   Economizer circuit

is sufficient to pick up the relay, the force will be far more than necessary to hold the relay in the picked-up position. To minimize heating of the coil, some dc contactors are arranged so that a resistor is inserted in series with the coil once the contactor has picked up, as shown in Figure 10-10. This current-reducing scheme is known as an economizer. Rather than use an external resistor, some manufacturers use additional turns of wire on the magnet coil to get the same effect.

Most ac motor starters use ac electromagnets to operate their contactors. When the contactor picks up, the air gap in the magnet structure closes, which increases the inductance (and therefore the inductive reactance) of the coil. So when the contactor picks up, the coil current is automatically reduced[11] and there is no need for an economizer circuit. Some large ac contactors use an ac control circuit, but their actuating electromagnet is a dc type fed from a full-wave rectifier. These circuits usually include an economizer of some kind.

DIRECT-CURRENT MOTOR STARTERS AND CONTROLLERS

The basic objective of a dc motor starter is to keep the starting current down to an acceptable value. If thrown directly on the line, a dc motor will draw 10 or 12 times

its full-load current, and such high currents

[11]If the contactor does not pick up properly (e.g., dirt is in the air gap of the magnet), the coil will usually burn out due to excessive current.

are likely to damage the commutator and brushes. There is also the question of whether the power supply system can carry such high currents without damage, and in a few cases the starting current may have to be limited in order to limit the motor torque.

The fuse or circuit breaker in a dc motor branch circuit is usually not larger than 150% of the motor full-load current. The starter has sufficient resistance to keep the starting current down to about 200%.

Manual DC Motor Starters

Three-Point Starters

Figure 10-11 shows a typical three-point starter. This starter is spring loaded to go to the *off* position and is held in the *run* position by an electromagnet. To start the motor, the operator should move the starter to the first contact and allow the motor to accelerate to as high a speed as it will reach. Then smoothly (and not too quickly) the starter arm is moved to the run position. To stop the motor, push the arm away from the run position and allow the spring to return it to the *off* position.

A three-point starter can be applied to shunt or compound motors and will prevent

these motors from running away owing to an open shunt field. However, if we attempt to get speed control using a shunt field rheostat, the starter tends to trip out when we raise the motor speed. Obviously, a three-point starter cannot be used with a series motor.

Occasionally, an operator will attempt to obtain a reduced motor speed by holding the starter in some intermediate position. This works, but the resistors are usually too small (physically) to dissipate the heat, and they will burn out if left in the circuit too long.

Four-Point Starter

As shown in Figure 10-12, the four-point starter has its holding magnet (usually with a series resistor) connected across the line, but it is otherwise identical to a three-point starter. This arrangement of the holding magnet makes the starter usable with series motors and also more suitable for shunt or compound motors when a field rheostat is to be used for speed control.

If the shunt field opens, a four-point starter cannot prevent the motor from running away. However, if the line fuses are rated no higher than 150% of the motor

FIGURE 10-11 Three-point starter

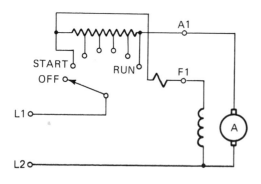

FIGURE 10-12 Four-point starter

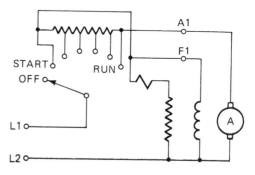

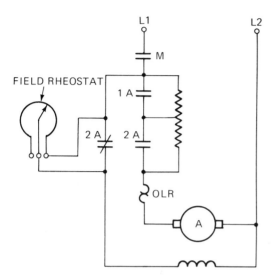

FIGURE 10-13   Typical power circuit for a dc starter

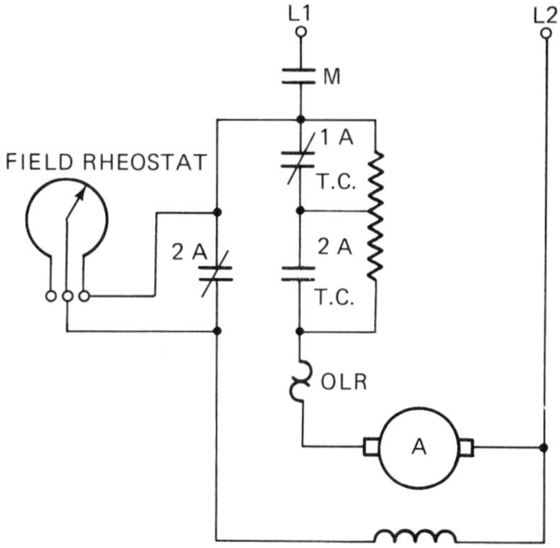

FIGURE 10-14   Alternative power circuit for a dc starter

full-load current, the fuses will usually blow before the motor reaches a dangerous speed. An overspeed switch can be wired into the circuit if desired.

Automatic DC Starters

Typical Power Circuits

Figure 10-13 shows a typical power circuit arrangement used for automatic dc starters using three contactors. The required cycle of operation is that *M* picks up (allowing the motor to start with all available resistance in series with the armature); after a suitable time delay 1*A* picks up, and after a further time delay 2*A* picks up. This particular circuit cuts out the starting resistor in two steps; but some starters do it in only one step, and there is no reason why we cannot use three or more steps if smooth acceleration is important.

The normally closed contacts on 2*A* are optional. If used, these contacts ensure that the motor will have maximum field flux during the starting period, so that it will develop maximum torque per line ampere.

An alternative power circuit arrangement is shown in Figure 10-14. The required cycle of operation is that 1*A* and 2*A* must pick up first, quickly followed by *M*. After a suitable time delay, 1*A* releases, and after a further time delay 2*A* releases. From the motor's point of view, these two power circuits do exactly the same thing. However, because the starters go through slightly different cycles, the control circuits are slightly different.

Typical Control Circuits

The control circuits for dc starters all tend to be quite similar. The most important variation is the method used to determine when the individual parts of the starting resistor should be shorted out. The following are three common techniques:

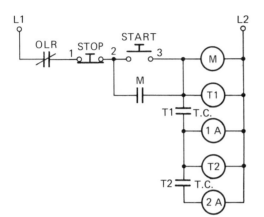

FIGURE 10-15 Control circuit for a definite-time
starter

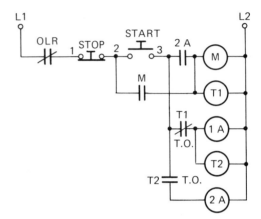

FIGURE 10-17 Another control circuit for a
definite-time starter

DEFINITE TIME STARTERS For a definite
time starter using the power circuit in Figure
10-13, we need a control circuit similar to
that shown in Figure 10-15. The time delays
of relays $T1$ and $T2$ are usually independently adjustable so that we can set them
to match the needs of the installation. Note
that $T1$ and $T2$ are both on-delay relays.

Instead of using separate time-delay relays,
some manufacturers use timed closing contacts on contactors M and $1A$ to replace

$T1$ and $T2$, respectively. The circuit then
reduces to that shown in Figure 10-16.

If we use the alternative power circuit
in Figure 10-14, the control circuit will resemble that shown in Figure 10-17. Note
that $T1$ is an on-delay and $T2$ is an off-
delay relay.

If contactors $1A$ and $2A$ are slow-release
relays, $T1$ and $T2$ in Figure 10-17 can be
eliminated, and the circuit in Figure 10-18
results. The timers can also be eliminated

FIGURE 10-16 Simplified control circuit for a
definite-time starter

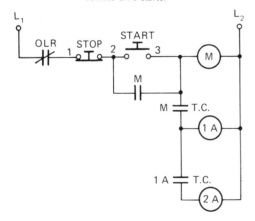

FIGURE 10-18 Control circuit using slow-release
contacts

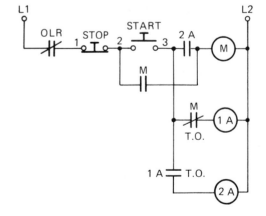

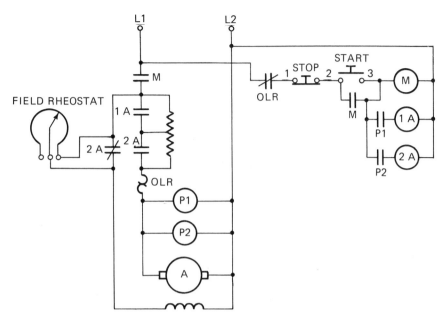

FIGURE 10-19   Typical counter-emf starter

by providing timed opening contacts on *M* and 1*A*.

COUNTER-EMF STARTERS   As a dc motor accelerates, the counter emf generated in the armature reduces the current flow through it. Reduced current means less voltage drop in the starting resistors, and therefore the voltage across the armature increases; in fact, the voltage observed at the brushes is only slightly greater than the counter emf.

A counter-emf starter has relays that sense the voltage across the armature and initiate the action of shorting out the starting resistors. Figure 10-19 is a typical counter-emf starter, and for a 125-V dc motor, *P*1 is adjusted to pick up with about 60 V and *P*2 picks up at about 90 V. This control circuit can also be adapted for use with the power circuit in Figure 10-14.

CURRENT-LIMIT STARTERS   A current-limit starter has one or more relays that monitor the armature current during the starting period. As the motor accelerates, the armature current goes down; when it reaches a low enough value, the current-sensing relays initiate the action of shorting out the starting resistor. Figure 10-20 is a schematic diagram of a current-limit starter. It is important to note that the current relays *S*1 and *S*2 pick up much more quickly than contactors 1*A* and 2*A*. Normally, *S*1 and *S*2 are adjusted to pick up at about the same current.

Reversing a DC Motor

We generally obtain reversed rotation by interchanging the connections to the armature, leaving the current in the field coils unchanged. In Figure 10-21, contactor *M*

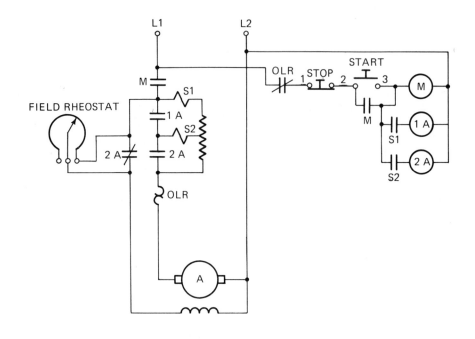

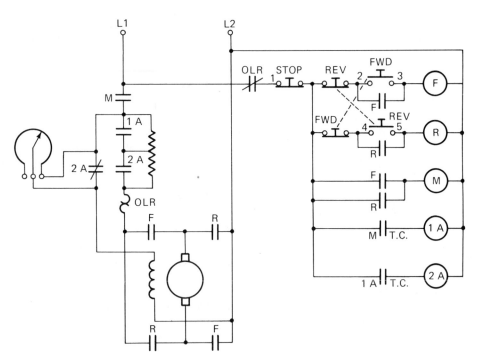

FIGURE 10-20 (opposite, top)   Current-limit starter

FIGURE 10-21 (opposite, bottom)   Reversing
controller for a dc motor

Dynamic Braking

serves as a start–stop switch, and contactors *F* and *R* determine the direction of rotation. Accelerating contactors *1A* and *2A* are operated from timed closing contacts on *M* and *1A*, respectively, but counter-emf or current-limit control would work as well. The circuit in Figure 10-21 permits the operator to plug-stop[12] the motor and for this reason is not always acceptable.

To prevent plugging, some manufacturers omit contactor *M* (so the shunt field is energized continuously) and add an antiplugging relay (*AP*), ending up with a circuit like that in Figure 10-22.

Dynamic braking is the practice of stopping a motor and its load by connecting the motor so that it will act as a generator, dissipating the electrical energy it produces. The simplest dynamic braking arrangements leave the shunt field of the motor energized all the time. To get braking effect, disconnect the armature from the line and put a resistor across the armature. Since the field is energized, the armature will circulate current through the resistor, dissipating the kinetic energy of the motor and its load.

[12]Plug-stopping is the practice of stopping a motor and its load by connecting the motor so that it develops torque in the direction opposite to its rotation.

FIGURE 10-22   Reversing controller with an antiplugging feature

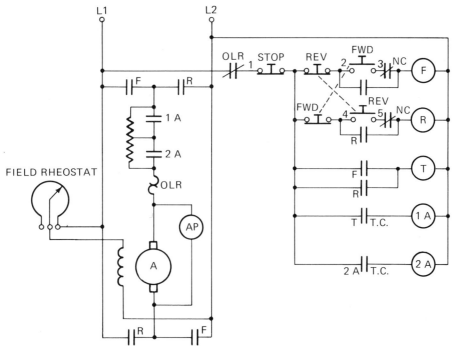

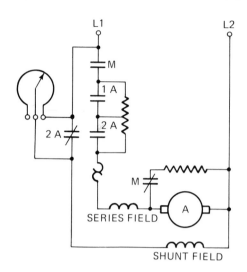

FIGURE 10-23   Elementary dynamic braking circuit

Figure 10-23 shows an elementary dynamic braking circuit. Note that the series field does not form part of the dynamic braking circuit. This is important, because if the motor is a cumulative compound type, it would become a differential compound generator, and this would seriously reduce the braking effect obtained.

Speed Control for DC Motors

Since there are only three basic ways to control the speed of a dc motor, let us consider each in turn.

Armature Circuit Resistance Control

To get speed control this way, we can use power circuits similar to those in Figures 10-11 through 10-14, but two modifications will be required. First, the resistors must be physically large enough to dissipate the heat produced when they are left in the circuit continuously. Second, the switching equipment must be such that all or part of

the starting resistance can be left in the circuit continuously. In some cases we may need more resistance (in ohms) to get the desired range of speed control than was needed just for starting purposes.

For manual operation, a drum controller can be used. Its switching operation is essentially the same as a three-point starter, but it will stay in whatever position it is placed and therefore has no holding magnet.

If the switching operations are to be done with contactors, the circuit in Figure 10-24 can provide three operating speeds. Notice that this is the same as the power circuit in Figure 10-13 plus the control circuit in Figure 10-15 with two sets of contacts (S_1 and S_2) added to the control circuit to block its operation. With S_1 and S_2 open, the motor runs at its lowest speed. With S_1 closed, it can progress to its second speed, and with S_1 and S_2 both closed it runs at its highest

FIGURE 10-24   Speed control by means of armature circuit resistance

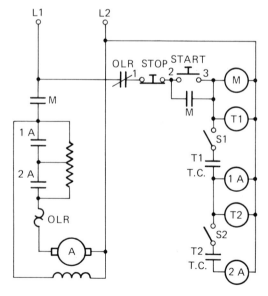

speed. Because of poor efficiency and poor speed regulation, armature circuit resistance control is losing popularity.

Field Flux Control

Speed control of shunt and compound motors is commonly obtained by using a manual rheostat in series with the shunt field, but this system does not lend itself to closed-loop control. It is possible to insert a transistor in series with the shunt field and get speed control by adjusting the base current to the transistor. We can also use circuits like those in Figure 10-28 to supply the field current. All these arrangements are adaptable to control from some external electrical signal.

Field flux control for series-wound motors can be obtained in several ways. We can permanently raise the speed by permanently connecting a low-resistance diverter in parallel with the field coils. The motor efficiency is not affected. If suitable taps have been provided, we can obtain speed control using a circuit like that in Figure 10-25. Reducing the number of turns used in the field coils

FIGURE 10-25   Speed control with tapped field coils

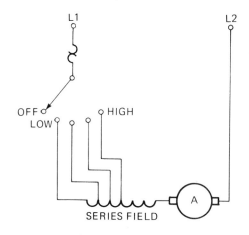

SERIES FIELD

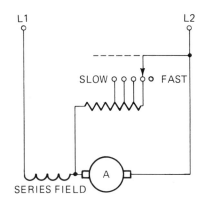

FIGURE 10-26   Reducing the speed of a series-wound motor

will raise the motor speed with no sacrifice of efficiency.

The third possibility is to switch in a resistor in parallel with the armature, as shown in Figure 10-26. This increases the field current and lowers the motor speed, but it is a rather inefficient arrangement.

Armature Voltage Control

For most installations, we obtain power for our dc motors by rectifying the available ac supply. To get a standard dc voltage (125 or 250 V), it is usually necessary to insert a transformer immediately ahead of the rectifier, but we will omit it from our diagrams. Figure 10-27 shows three common rectifier arrangements using solid-state diodes.

To get an adjustable dc voltage, we substitute SCR's for some or all of the diodes in Figure 10-27 and end up with the circuits in Figure 10-28. By adjusting the timing of the gate pulses, we can adjust the output voltage from maximum down to zero. Good efficiency and smooth, wide-range speed control can be obtained with these systems.

For motor-speed control we use an adjustable-voltage power supply to feed the armature, and if the motor has a shunt field,

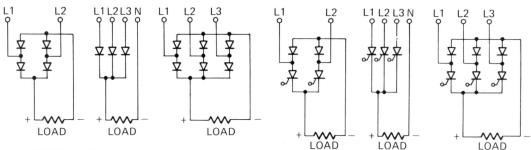

FIGURE 10-27   Some common rectifier circuits

FIGURE 10-28   Adjustable-voltage dc power supplies

that field will be fed from a separate power supply, which may or may not be adjustable. To get low motor speeds, we need maximum field current and minimum armature voltage; for high-speed operation we need maximum armature voltage and minimum field current, and we cannot get that result using only one power source.

The reader is hereby cautioned not to use a rectifier type of power supply to feed a dc motor unless the motor is designed for it. Whether the output voltage is adjustable or not, all rectifier-type power supplies have a more or less pulsating output, and this ripple, as it is called, causes additional iron losses (and therefore heating) in the motor.

STARTERS AND CONTROLLERS FOR THREE-PHASE INDUCTION MOTORS

Line Starters for Squirrel-Cage Motors

In contrast with dc motors, ac motors can generally be started with full rated voltage applied to them during the starting period. A starter designed to do this is known as a line starter. There are three basic types:

Manual Starters

Suitable for motors up to about 6 kW at 600 V, these starters are three-pole manual switches with an overload relay on each pole. If any one overload relay trips, the mechanical linkage is such that all three poles of the switch will open.

Magnetic Line Starters

These are magnetic contactors equipped with a three-pole overload relay and are used for motors of up to several thousand kilowatts. The overload relay contacts are wired in series with the actuating magnet coil, and if any one pole of the overload relay trips, the magnetic contactor will release. The external control circuit is the same as that used for a dc starter and/or closely resembles the circuits in Figure 10-4 and 10-6, whichever is required. If the line voltage is not more than 600 V, the control circuit may be supplied directly from the line, but for higher voltages a control transformer is required. If a control transformer is used (and they are quite common even on 240-, 480-, and 600-V equipment), the control circuit will generally operate at 120 V.

A three-phase motor is reversed by interchanging any two of the three input lines. A reversing starter uses two three-pole contactors to achieve the desired results, as shown in Figure 10-29. The contactors are usually interlocked mechanically and electrically so that they cannot both be in the picked-up position at the same time.

Solid-State Starters

Figure 10-30 shows the power circuits for two solid-state motor starters. The triggering circuits (not shown) are designed to turn

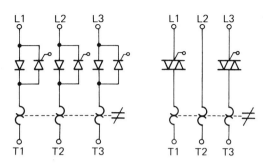

FIGURE 10-30 Two solid-state motor starters

on the SCR's or TRIAC's very early in the cycle. A standard two- or three-wire external control circuit is used to start and stop the triggering circuits, thereby controlling the motor. For their own safety, maintenance electricians should note that these starters do not isolate the motor from the line.

Reduced-Voltage Starters for Squirrel-Cage Motors

Any scheme that will reduce the starting current to a motor will reduce the voltage dip produced on starting, and this is the main reason why the variety of reduced-voltage starting methods that have been developed are so widely used. But the name "reduced voltage" is slightly misleading. Motor starting currents can be reduced by using a step-down autotransformer or by starting with a series impedance, or by reconnecting the stator winding of the motor to give it more impedance, and all these techniques are known as reduced-voltage starting methods. Reduced-current starting would be a better name, because the last two methods do not involve any reduction of voltage.

Reduced-voltage starting has one unfortunate side effect: the torque produced by the motor is proportional to the square of

FIGURE 10-29 Reversing and nonreversing line starters

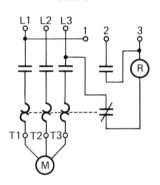

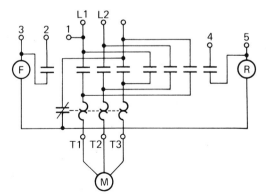

the voltage applied and is therefore drastically reduced. Since the current drawn tends to remain at a high value until the motor approaches full speed, the motor must be able to start and accelerate its load to almost full speed with reduced voltage in order to obtain any real reduction of line current. It therefore follows that the torque requirements of the load set a lower limit on the minimum possible motor starting current.

Consider a load, motor, and starter combination with torque and current characteristics as shown in Figure 10-31, and note that the motor torque on reduced voltage exceeds the load torque all the way up to about 90% of synchronous speed (point *B*). The best practice with this combination is to apply reduced voltage to the motor and allow it to accelerate to the maximum possible speed, then change over to full voltage. The motor torque and current will then follow the curves *ABCD* and *PQRS*, respectively, and assuming closed transition,[13] the line current and line voltage variations that result are shown in Figure 10-32.

Changing to full voltage before the motor has reached point *B* will cause "more than minimal" starting current to flow, sometimes approaching the "locked rotor and rated voltage" values. This may blow the fuses in the motor branch circuit and always creates an unnecessary voltage dip. Automatic motor starters usually have a timer that can be set to permit adequate starting time; but where manual reduced-voltage starters are used, the operators frequently need instruction on this point.

---

[13]"Closed transition" means that the motor circuit is not interrupted when changing over to full voltage. Starters that interrupt the motor current at changeover time are known as "open-transition" types.

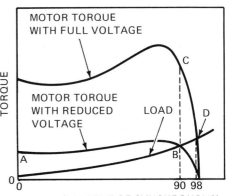

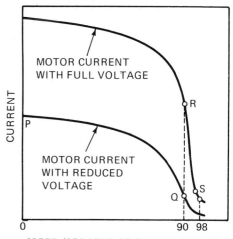

FIGURE 10-31 Torque and current characteristics of a load and motor (usual case)

FIGURE 10-32 Line voltage and motor starting current with closed transition

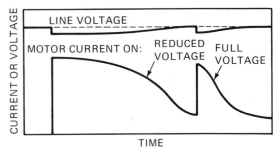

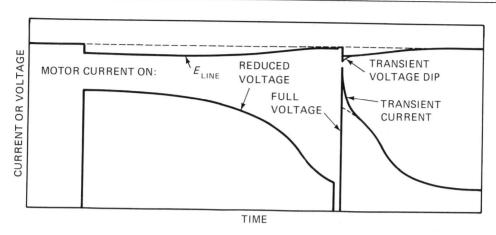

FIGURE 10-33 Line voltage and motor starting current with open transition

If a reduced-voltage starter has open transition, the line current and voltage variations tend to be more like the graph in Figure 10-33. The brief (5 or 6 cycle) transient current that may flow at changeover time can cause objectionable light flicker. Closed-transition starters are therefore considered preferable, but open-transition equipment is by no means rare.

Consider a load, motor, and starter combination having torque and current characteristics as shown in Figure 10-34, and note that on reduced voltage the starting torque of the motor is less than that required by the load. In this case it is obvious that the motor will not start on reduced voltage; so when the circuit changes over to full voltage, the usual "line start" current will flow. We have here a case of increment starting; that is, when initially energized the motor turns very slowly or not at all, but the starting current rises in steps or increments, as shown in Figure 10-35, until the motor torque becomes sufficient to accelerate the load to practically full speed.

FIGURE 10-34 (right) Torque and current
characteristics of a load and motor
(increment starting)

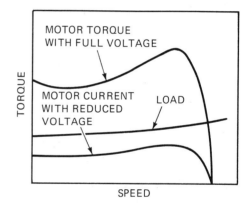

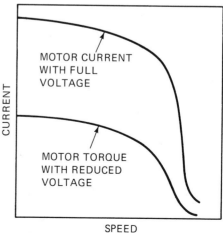

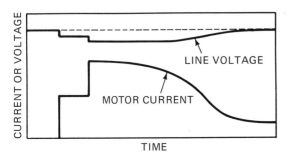

FIGURE 10-35 Line voltage and current with
increment starting

step should therefore be minimal in such
a case.

Autotransformer Starters
(Starting Compensators)

Figure 10-36(a) and (b) shows two common power circuit arrangements for these

With two-step[14] increment starting, the
voltage drop appears on the line in two small
dips rather than one large one, making light
flicker less noticeable, and this is its main
advantage. But if the supply system voltage
regulators can compensate for the first voltage dip before the second one occurs, the
total voltage dip will be reduced accordingly.

Increment starting is practical only with
closed-transition starters. If attempted with
open-transition equipment, increment starting will cause voltage flicker comparable to
line voltage starting equipment, and the latter
is a good deal less expensive.

The two-step reduced-voltage starters are
most common, but three or more steps are
sometimes used. With starters having three
or more steps, increment starting (if it occurs) is generally limited to the first step.

Reduced-voltage starting does not greatly
affect motor temperature rise during the starting period. While the rate of temperature
rise is decreased, acceleration times are increased, and the final temperature remains
about the same.

If increment starting is involved, motor
heating occurs on the first step, but acceleration does not. The time spent in the first

FIGURE 10-36 Power circuits for autotransformer
starters

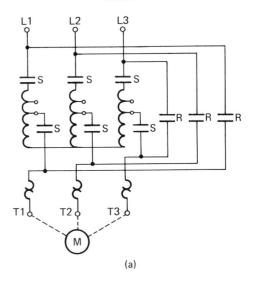

(a)

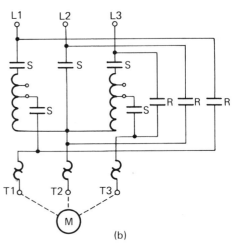

(b)

[14]The number of steps includes each reduced voltage
step and also the final line voltage step.

devices, and whether manual or automatic the cycle of operation must be as follows: the contacts marked S must close first, and after the motor has accelerated to maximum speed, these contacts must open and the contacts marked R must close.

Manual compensators are usually located near the motor of the driven machine so that the operator can tell when to change over to full voltage. Automatic types usually have a time-delay device that initiates the changeover and that must be field adjusted to allow the necessary accelerating time. The schematic diagram in Figure 10-37 is a typical control circuit for an automatic compensator.

Preferred standard voltage taps for autotransformer starters are 80% and 65% for all sizes, with an additional 50% tap on larger starters. The voltage across the motor remains constant throughout the starting period, and so the motor torque throughout the speed range will be 64%, 42%, or 25% of that which would be obtained from the same motor if controlled by a line starter.

While the motor starting current is always proportional to the voltage tap used, the turns ratio of the autotransformer reduces the starting line current[15] even further. With

[15]This is the current on the supply side of the starter.

preferred standard taps, the starting line current is reduced to about 64%, 42%, or 25% of normal.[16] Autotransformer starters and wye–delta starters are the only ones that reduce current and starting torque in the same proportion. With all other types, the loss of starting torque in percent will always be greater than the reduction of starting current in percent.

To minimize the starting current, a compensator should be set on the lowest tap at which acceleration of the load is satisfactory. Autotransformers are usually designed for short-time duty only. They may overheat if the motor is started too frequently.

The simplest designs of starting compensators are inherently open-transition devices and may cause voltage flicker at changeover time. But the Korndorffer circuit overcomes this limitation. The power circuit arrangement for the Korndorffer scheme is shown in Figure 10-38(a) and a typical control circuit schematic in 10-38(b). The cycle of operation is that contactors S and W pick up, and after a suitable time delay, W releases and R picks up. These two latter operations occur in quick succession.

Primary Resistor Starters

The power circuit arrangement and typical control circuit for a primary resistor starter are shown in Figure 10-39(a) and (b). The cycle of operation is that contactor M picks up first, and after a suitable time delay, contactor $1A$ picks up.

The resistors used may be nichrome wires or strips, cast iron or bronze grids, or graphite discs stacked together; they are usually designed for intermittent duty only. If the

[16]These percentages do not include magnetizing current for the autotransformer. The actual starting line currents will be slightly higher.

FIGURE 10-37   Control circuits for autotransformer starters

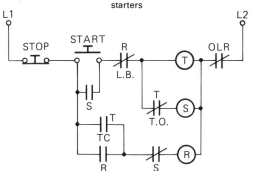

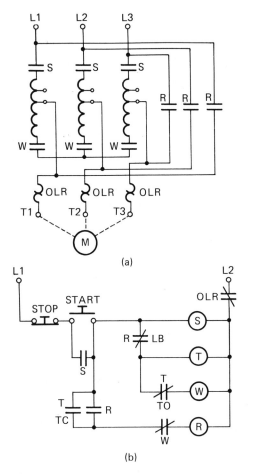

(a)

(b)

FIGURE 10-38   Autotransformer starter with closed
transition

its normal starting torque. However, as the motor accelerates, its internal impedance becomes greater and more resistive in character, and this increases the voltage across the motor, as shown by the phasor diagram in Figure 10-40. The starting torque is therefore the same, but as the motor approaches full speed, the torque becomes greater than would be developed by the same motor fed from an autotransformer starter set to give the same initial voltage.

This comparison is illustrated graphically in Figure 10-44. Primary resistor starters are inherently closed-transition devices and are therefore suitable for use as increment

FIGURE 10-39   Two-step primary resistor starter

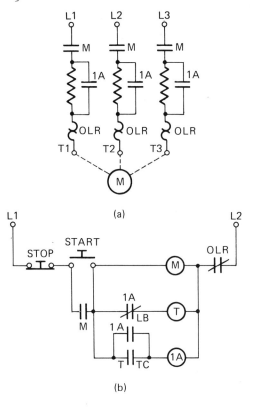

(a)

(b)

starter operates too many times per hour, or if contactor 1*A* fails to pick up, resistor damage will usually occur. Thermal protection for the resistors can be included in the control circuit if desired.

The resistors used in these starters are usually selected and/or adjusted so that the initial voltage across the motor will be either 80% or 65% of normal. The motor then draws either 80% or 65% of its normal starting current and develops either 64% or 42% of

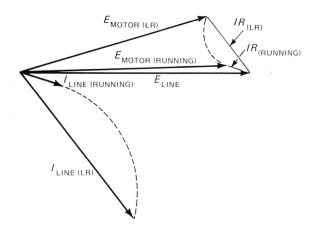

FIGURE 10-40   Phasor diagram for resistance starters

starters. To obtain smooth starts and minimize voltage disturbances, the starting resistors may be cut out in two or more steps or graphite compression-type resistors may be employed.

Primary Reactor Starters

As Figure 10-41 shows, these starters are similar to primary resistor starters except that reactance coils are used to limit the starting current. The cycle of operation and the current and torque reductions are the same as for resistor starters, but the reactance starter is used with large motors where heat dissipation from starting resistors would be a severe problem.

Although reactors reduce starting current, they also reduce power factor on starting. If the supply system impedance is mostly inductive reactance and starting reactors are used, the angle between starting current and voltage tends to approach the impedance angle of the supply system, and this increases line drop as shown in Figure 10-42.

FIGURE 10-41   Power circuit schematic for a reactance starter

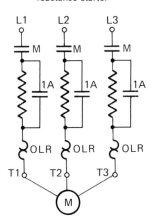

FIGURE 10-42   How power factor affects line drop

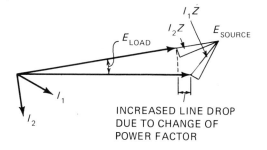

INCREASED LINE DROP
DUE TO CHANGE OF
POWER FACTOR

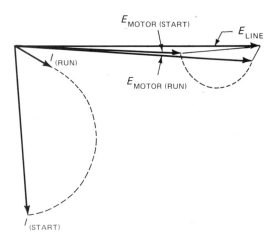

FIGURE 10-43   Phasor diagram for reactance
starting

As a result, the reduction of voltage dip is sometimes not as great as the reduction of starting current.

As the motor accelerates, the voltage across it increases as shown in Figure 10-43, and the torque increases as shown in Figure 10-44. These changes are more pronounced with reactor starters than they are with resistance types.

FIGURE 10-44   Effect of reduced-voltage starters
on motor torque

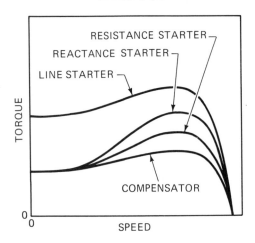

Heating of the reactors usually limits the permissible number of starts per hour. These starters are inherently closed transition, which makes increment starting practical.

Solid-State Reduced-Voltage Starters

These starters use power circuits like those in Figure 10-29. During starting the SCR's or TRIAC's are triggered late in the cycle and then gradually advanced toward the beginning. This gives the effect of gradually increasing the voltage on the motor, similar to resistor or reactor starting.

Part-Winding Starters

Part-winding starting is accomplished by winding the stator of the motor in two parallel paths (either wye or delta) with separate leads for each path. Each path is then energized through a separate magnetic switch. Some possible configurations are shown in Figures 10-45(a) and (b) and 10-46(a). A typical control circuit is shown in Figure 10-46(b), and the cycle of operation is that contactor S picks up first and after a suitable time delay contactor R picks up. The permissible number of operations per hour is limited only by heating of the motor.

The two parallel paths in the motor winding do not have to be, but usually are, identical, and one might then expect the starting current to be cut in half. But if both paths are energized, mutual inductance between the windings creates an additional reactance, which does not exist with only one part of the winding energized. So the starting current turns out to be about 65%, and the torque is around 42% of normal.

Field adjustment of the motor starting current and starting torque is not possible on part-winding starters. The torque is so low that increment starting may be unavoid-

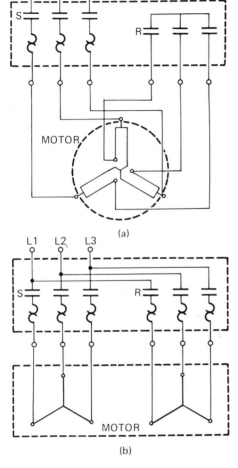

(a)

(b)

FIGURE 10-45   Some power circuit schematics for part-winding starters

given size of motor the overload relays must be smaller if two sets are used, but the starting current remains the same. It therefore becomes more difficult to select fuses that will accommodate the starting current and still protect the overload relays against short-circuit currents.

A part-winding starter should only be applied to a motor specifically designed for this purpose. It is easy to connect an ordinary dual-voltage motor to a part-winding starter on a 230-V system. But with only half the winding energized, many of these

FIGURE 10-46   Power and control circuit for a part-winding starter

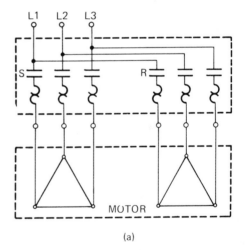

(a)

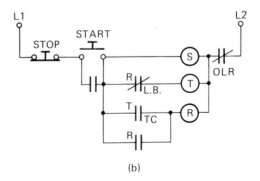

(b)

able with some kinds of loads, but the starters have closed transition, so such an installation might still be practical.

It should be noted that some of these starters have a separate set of overload relays for each half of the motor winding. This is desirable because a single set of overload relays, as shown in Figure 10-45(a), cannot protect the motor against burnout if contactor *R* fails to pick up. However, for a

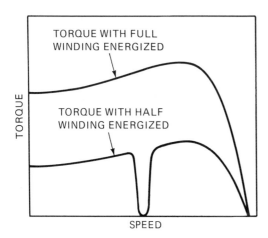

FIGURE 10-47   Possible torque curves for a dual-voltage motor

motors have a severe torque dip at about half-speed, as shown in Figure 10-47. If this is the case, increment starting will be unavoidable. Excessive heating and objectionable noise on starting may also occur.

Wye–Delta Starter

If a three-phase motor is designed to operate with the winding connected in a delta, and a separate lead is provided for each end of each phase, then a wye-start delta-run starter can be applied to it. A schematic diagram and a typical control circuit for such an arrangement are shown in Figure 10-48(a) and (b). The cycle of operation is that contactors M and S pick up together, and after a suitable time delay S releases and R picks up. The permissible number of starts per hour is limited only by heating of the motor. If taken to a repair shop, any delta-connected motor can be easily adapted to this scheme.

Wye–delta starters reduce both the starting current and the motor torque to 33% of normal. Such low torque may be inadequate

for many loads, and increment starting will then be the result. These starters are inherently open-transition designs; this is a rather serious drawback because of the transient inrush current that flows at changeover time, and increment starting is not practical with open transition.

FIGURE 10-48   Power and control circuits for a wye-delta starter (open transition)

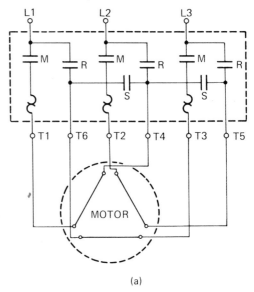

(a)

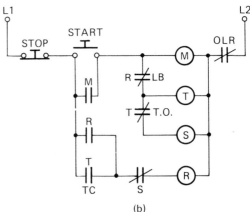

(b)

But wye–delta starters with closed transition are available. Schematic diagrams of the power and control circuits of such a starter are shown in Figure 10-49(a) and (b). The cycle of operation is that contactors M and S_1 pick up together, and after a suitable time delay S_2 picks up, S_1 releases, and R picks up, in that order. These last three operations occur very quickly. Increment starting is practical with closed-transition wye–delta starters, but it is seldom done.

FIGURE 10-49   Power and control circuits for a wye-delta starter (closed transition)

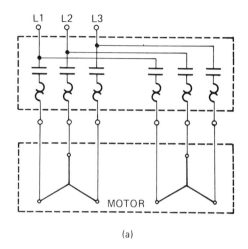

(a)

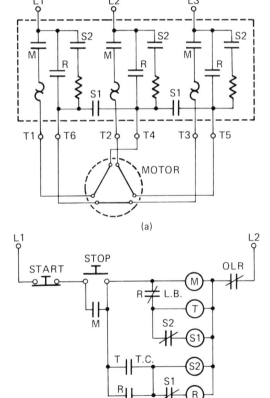

(a)

(b)

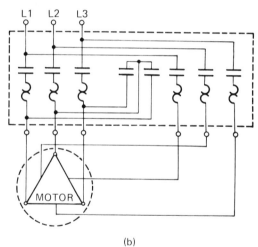

(b)

FIGURE 10-50   Power circuits for two-speed motor controllers. (figure continued on next page)

Controllers for Multispeed Motors

Two-speed two-winding motors can be controlled by two three-pole contactors as shown in Figure 10-50(a). Two-speed single-winding motors require one three-pole and one five-pole contactor, as shown in Figure 10-50(b), (c), and (d). The contactors must

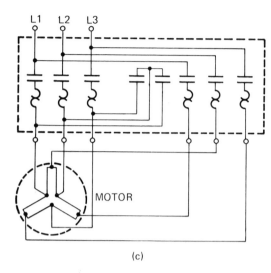

(c)

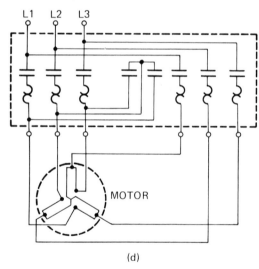

(d)

FIGURE 10-50 *(cont.)*

be mechanically interlocked to prevent short circuits or to prevent both windings from being energized at the same time. Two sets of overload relays are required, one set for each speed. The control circuit for a two-speed controller is the same as that used for a reversing controller.

Three- and four-speed controllers are similar to those required for two-speed single-winding motors. Although the switching patterns are the same, the relative power load switched by each contactor depends upon whether a constant-power, constant-torque, or variable-torque motor is being controlled. Multispeed controllers are designed with this fact in mind and must not be indiscriminantly applied or wrongly connected to the motor. Care must also be taken to obtain the desired direction of rotation at each speed.

If a two-speed controller is applied to a constant-torque or variable-torque motor, it becomes difficult to select fuses that will accommodate the high starting current of the motor on its high-speed connection and still protect the small overload relays required for low speed.

Two-, three-, and four-speed controllers are also available with provision for reduced voltage starting if the latter feature is desired.

Starters and Controllers for Wound-Rotor Motors

Manual Starters and Controllers

These use a line starter to switch the stator circuit, and reduced voltage starting is never employed. The face plate design is commonly used for small starters and controllers. A typical diagram is shown in Figure 10-51. Note that the resistance is cut out of all three phases simultaneously so that the rotor circuit is balanced at all times.

The resistance steps, however, are usually not uniform. The first resistance steps to be

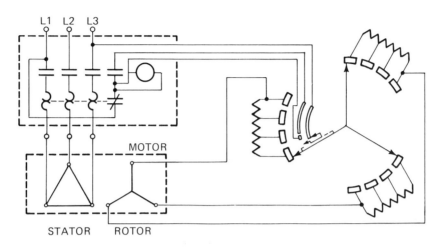

FIGURE 10-51 Face-plate starter

cut out (as the starter is moved to the run position) are usually the largest, and the last resistance steps are usually the smallest. Notice, too, that the pilot contacts are connected much like a stop–start push-button station and will prevent starting the motor with reduced rotor resistance, which would allow excessive current to flow.

If used for speed control, the resistors must be designed for continuous duty and may require greater ohmic value than that required only for starting.

The drum-switch design shown in Figure 10-52 is also used for either starting or speed control. Notice that this device single phases the rotor circuit when placed in the first position. This reduces the motor torque considerably and therefore reduces the mechanical shock applied to the drive system on starting. But the motor will usually be quite noisy while in this condition. Advancing the controller to step 2 closes the remaining phase of the rotor circuit, and moving the controller beyond that point progressively

short circuits the resistors. From the ascending spiral pattern of the terminal numbers on the resistors, it can be seen that resistance is cut out of the circuit one section at a time from each phase in sequence. At many controller positions, then, the rotor circuit will be unbalanced. This may cause a slight magnetic noise in the motor, but otherwise is not important.

The reversing drum controller shown in Figure 10-53 may be intended either for starting duty only or speed control. Notice the use of the auxiliary contacts in the control circuit of the line starter, as well as a push-button station. Notice, too, that the auxiliary contacts at the top of the drum switch serve to reverse the phase sequence of the voltage applied to the stator, and that the rotor connections need not be changed to reverse rotation.

Automatic Starters and Controllers

Like the manual types, these use a line

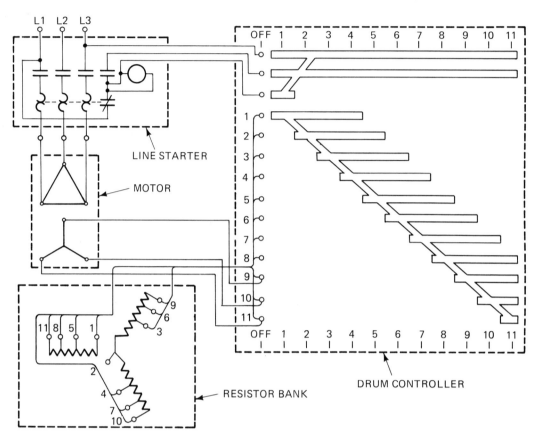

FIGURE 10-52   Nonreversing drum controller

starter to switch the stator circuit, but they use additional contactors to switch the rotor resistance out as required. Figure 10-54 shows a typical power circuit arrangement. For starting purposes, the required cycle of operation is that M must pick up first. After a suitable time delay, $1A$ picks up, and after a further time delay $2A$ picks up.

A control circuit that will provide this cycle is shown in Figure 10-55. To get speed control, add switches to the control circuit that will prevent $2A$ and possibly $1A$ from picking up. Three-speed operation can be obtained.

Figure 10-56 shows an alternative way of connecting the accelerating contactors $1A$ and $2A$ to the rotor resistors. The advantage of this is that if the two sections of resistance in each phase are not equal we can obtain four-speed operation instead of three, but a more elaborate control circuit will be required.

FIGURE 10-53 (opposite, top)   Reversing drum controller

FIGURE 10-54 (opposite, bottom left)   Power circuit for a wound-rotor motor controller

FIGURE 10-55 (opposite, bottom right)   Control circuit for a wound-rotor motor controller

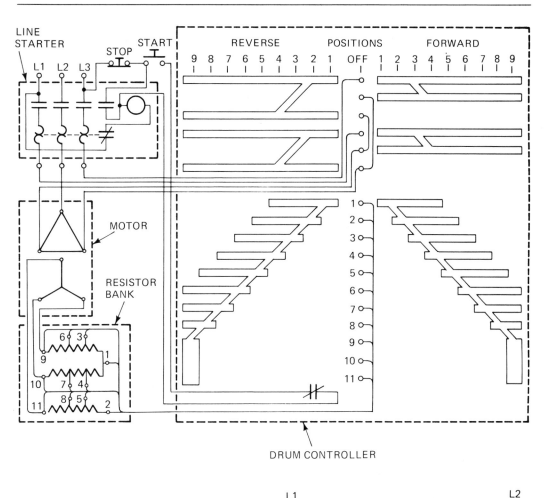

DRUM CONTROLLER

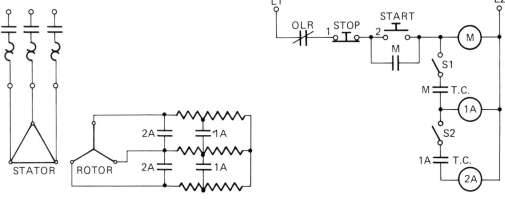

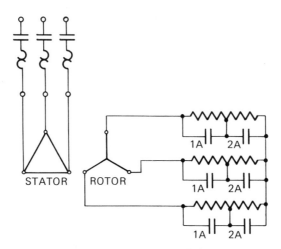

FIGURE 10-56   Alternative connections for the accelerating contactors

STARTERS AND CONTROLLERS FOR SINGLE-PHASE MOTORS

Manual starters may be either one or two pole but may have only one overload relay. Magnetic line starters are generally two pole with only one overload relay. They are basically three-phase line starters from which the unnecessary parts have been omitted. For reversing applications we often use a manual starter (to provide overload protection) plus a three-position switch (forward, off, reverse), which amounts to the four-pole double-throw switch shown in Figure 10-57. Two four-pole contactors (mechanically interlocked) could do the same thing. Only one overload relay is required.

It is important to realize that single-phase motors with centrifugally operated starting switches will not plug-stop. If the switch in Figure 10-57 is quickly moved from forward to reverse, the motor will continue to run in the forward direction, and in some situations this could be hazardous.

For two-speed motors, one could use two 2-pole contactors (suitably interlocked) and two overload relays, one for each speed. If manual control is acceptable, use a DPDT center-off switch plus two manual motor

FIGURE 10-57   Reversing a single-phase motor

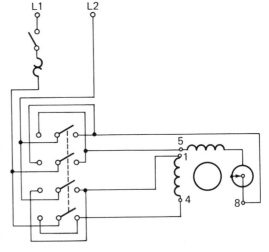

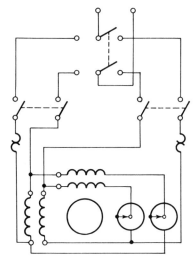

FIGURE 10-58   Manual control for a two-speed motor

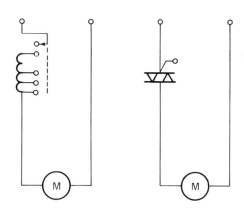

FIGURE 10-59   Speed control of shaded-pole motors

starters that will normally be left in the *on* position. Arranged as shown in Figure 10-58, this will provide proper overload protection for each speed.

For speed control of ordinary shaded-pole or permanent-split capacitor motors driving fan loads, use one of the circuits in Figure 10-59. Using more of the turns of wire in the choke coil or delaying the gate pulse to the TRIAC reduces the motor speed.

SYNCHRONOUS MOTOR STARTERS

The stator circuit of a synchronous motor is the same as that of a squirrel-cage motor, so we use the same kinds of starters and provide the same kind of overload protection. However, we need some additional equipment to switch the field circuit as required. To get maximum pull-in torque, field excitation must be applied only when the motor is running near synchronous speed, and then only when the rotor poles are in the correct position relative to the stator magnetic poles. To do this automatically, we must obtain an electrical signal indicating that the two conditions have been met. We must also protect the motor against failure to synchronize on start-up and/or loss of synchronism while running, because prolonged operation this way will overheat the squirrel cage. Again we will need an electrical signal indicating motor slip.

Slip Detection

If the motor is not synchronized, the relative motion between the rotor and the stator magnetic field induces an ac voltage in the rotor winding. The magnitude and frequency of this voltage will be directly proportional to the slip.

During the starting period, the field winding must be connected to a discharge resistor, and the easiest way to detect the slip is to

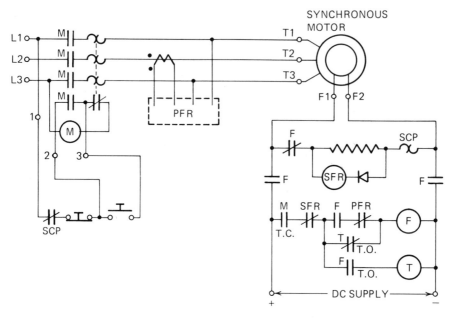

FIGURE 10-60 Synchronous motor starter (slip-frequency relay type)

monitor the voltage or current in the discharge circuit. The lower this voltage or current becomes, the more closely the motor has approached synchronous speed. A thermal overload relay in the field discharge circuit, as shown in Figures 10-60 and 10-61, can protect the motor from prolonged slippage with no field excitation.

If a motor is slipping with excitation applied, the induced rotor voltage will superimpose an ac current on top of the dc excitation current. We can isolate this ac component by inserting a transformer in the excitation circuit. If the transformer has an output voltage, the motor must be slipping.

Loss of synchronism can also be detected by monitoring the reactive power input to the stator. As the torque angle increases, the reactive power input becomes high and lagging. A relay that responds to this condition can detect the first slip cycle, and if connected

like relay *PFR* in Figure 10-60, excitation will be removed and then reapplied.

Rotor Pole Position Detectors

For maximum pull-in torque, field excitation should be applied when the flux linking the rotor and stator winding is at a maximum and in the same direction that the rotor mmf will be when the direct current is applied. This condition exists when the induced rotor voltage is zero and is just starting to build up in the positive direction (positive direction is the direction of normal excitation current flow).

Referring to Figure 10-60, the slip frequency relay (SFR) will give us the required action. Because of the diode, SFR receives only the positive alternations of the induced rotor voltage. However, it picks up, and because it needs about $\frac{1}{6}$ s to release, it remains

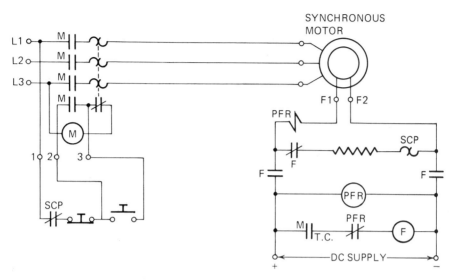

FIGURE 10-61 Synchronous motor starter (polarized-field relay type)

picked up as long as the rotor frequency is high. However, as the motor speeds up, the length of each alternation gradually increases. When the length of the negative alternation (when the relay is deenergized) becomes equal to the release time of the relay, SFR releases and excitation is then applied to the motor. Because the length of the negative alternation increases gradually, SFR will always release at or near the beginning of a positive alternation, which ensures that maximum pull-in torque will be produced.

The polarized field relay (PFR) in Figure 10-61 has a dc polarizing coil and a current coil connected in series with the motor field. If the current flow is the proper direction in the current coil, this relay picks up, and if the rotor frequency is high, it stays picked up. However, if the rotor frequency is low, the relay releases near the end of the negative alternation, which is the appropriate time to apply field excitation to the motor. If the motor loses synchronism while running, PFR in Figure 10-61 will automatically remove and reapply field excitation.

SUMMARY

Wherever you find a motor, you can expect to see a starter or controller associated with it. These starters and controllers perform the switching operations required to start and stop, reverse rotation and/or change the motor speed, and, in addition, protect the motor against excessive mechanical load.

The starters and controllers used with a given type of motor all have very similar power circuit arrangements. The actual switching operations may be accomplished either manually or by means of electromechanical (or solid state) contactors or by means of relays. The power circuit arrangement dictates the required cycle of switching operations.

Contactors and relays (be they conventional or solid state) are the fundamental building blocks used to construct automatic starters and controllers. In addition, they enable us to interface power equipment like motor starters with control equipment like logic circuits or computer systems, and therefore make it possible to assemble complete automatic controls for systems involving dozens or even hundreds of motors.

REVIEW QUESTIONS

10-1. What is an electromechanical relay?
10-2. What is the difference between a schematic and a wiring diagram?
10-3. In what position is a relay usually shown?
10-4. What is meant by the terms "pickup" and "release"?
10-5. What is a pilot device?
10-6. What are the advantages of relays?
10-7. What is meant by a two-wire control system?
10-8. Draw a diagram of a three-wire control circuit for a relay and explain step by step what happens when the following occur:
(a) The start button is pushed.
(b) The stop button is pushed.
(c) Both push buttons are pushed together.
10-9. What is meant by the term "electrically interlocked relays"?
10-10. Relay B has a start–stop push-button control circuit and is to be unilaterally interlocked with relay A, with B dependent on A. For each of the following requirements, specify (i) whether a normally open or normally closed contact from A is required in the control circuit for B, and (ii) how it should be connected (series or parallel with the existing turn on, turn off or holding sets of contacts for B).

(a) When A is picked up, B can be turned on normally but cannot be turned off until A is released.
(b) B cannot be turned on unless A is picked up. However, it will hold and/or can be turned off regardless of the status of A.
(c) B picks up whenever the start button is pressed but will not hold unless A is picked up.
(d) B picks up and holds whenever A is released. Pushing the stop button will momentarily release B, but it will not stay released unless A is picked up.
(e) B will not hold and cannot be initially picked up unless A is released.
10-11. Two relays A and B are to be symmetrically interlocked electrically so that neither relay will pick up or hold unless the other is released. Draw a schematic diagram to show how this can be done (provide a start button for each relay and use only one stop button for both).
10-12. What is meant by overlapped contacts?
10-13. Distinguish between a motor starter and a motor controller.
10-14. Distinguish between an on-delay and an off-delay timer.

10-15. Draw a diagram and carefully label the contacts of the following:
   (a) An off-delay timer that has two pairs of normally open time-delayed contacts and one pair of normally open instantaneous contacts.
   (b) An off-delay timer with one pair of timed opening contacts and one pair of timed closing contacts.

10-16. What four methods are used to provide overload protection for motors?

10-17. Which types of overload relays:
   (a) Cannot be arranged to reset automatically?
   (b) Are adversely affected by ambient temperature?
   (c) Can protect a motor against a higher than normal ambient temperature?

10-18. Draw schematic diagrams of a three-point and a four-point manual starter.

10-19. Draw a power circuit for a dc starter similar to the one in Figure 10-13 but using three accelerating contactors (1A, 2A, and 3A) rather than two. Be sure to label all the contacts shown.

10-20. Draw control circuit schematics similar to those in Figures 10-15 and 10-16, but suitable for use with the power circuit for question 10-19.

10-21. Referring to Figures 10-14, 10-17, and 10-18, repeat questions 10-19 and 10-20.

10-22. The power circuit in Figure 10-14 is to be used to build a counter-emf type of starter. Draw a suitable control circuit.

10-23. The power circuit in Figure 10-14 is to be used to build a current-limit starter. Draw a suitable control circuit.

10-24. Modify the control circuit in Figure 10-21 so that the motor cannot be reversed unless the operator first pushes the stop button.

10-25. Explain how the circuit in Figure 10-22 prevents the operator from plugging the motor.

10-26. Write out the cycle of operation of the circuit in Figure 10-23. Include both starting and stopping the motor.

10-27. Referring to the control circuits for questions 10-20 and 10-21, modify those four circuits, adding manually controlled contacts as necessary so that they will now act as armature-resistance-type controllers, giving four-speed operation of the motors they control.

10-28. There are several ways to adjust the field flux of a series-wound motor. What methods can be used and how does each one affect motor speed and efficiency?

10-29. Draw schematic diagrams of the power circuits of:
   (a) A single-phase full-wave bridge rectifier having fixed output voltage.
   (b) A three-phase half-wave rectifier with adjustable output voltage.

10-30. A dc motor runs cooler if powered from a battery or a generator rather than a rectifier. Why?

10-31. What is meant by the term "increment starting" and why does it occur?

10-32. What is meant by the term "line starter"?

10-33. Draw a diagram of a solid-state starter.

10-34. When turned off, does a solid-state starter isolate the motor from the line? Explain.

10-35. What is meant by "open transition," and why is "closed transition" considered preferable?

10-36. What is the main reason for the use of reduced-voltage starters with ac motors?

10-37. Draw schematic diagrams of the power circuits for each of the following motor starters and write out the cycle of operation:
   (a) Autotransformer starter (open transition).
   (b) Autotransformer starter (closed transition).
   (c) Primary resistor starter (two step).
   (d) Part-winding starter.
   (e) Wye–delta starter (open transition).
   (f) Wye–delta starter (closed transition).
   (g) Solid-state reduced-voltage starter.

10-38. A motor provides more starting torque per line ampere with an autotransformer starter than it does with a primary resistance starter. Why?

10-39. Why are reactors used for starting large motors rather than resistors?

10-40. The circuit in Figure 10-45(a) is a poor arrangement for a part-winding starter. Why is this circuit inferior to the others?

10-41. Why should one not arbitrarily use a dual-voltage wye-connected motor for part-winding starting?

10-42. When selecting a controller for a two-speed motor, why must one know:
   (a) Whether the motor has one or two stator windings?
   (b) Whether the motor is a constant-power, constant-torque, or variable-torque design?

10-43. Starters for wound-rotor motors are designed so that the stator cannot be initially energized unless the rotor circuit resistance is at maximum. Why?

10-44. What is the advantage of single phasing the rotor circuit on the first step of a wound-rotor motor starter?

10-45. Referring to Figure 10-55, carefully explain how this circuit can provide four motor speeds.

10-46. Draw a diagram showing two 4-pole contactors arranged to reverse a single-phase motor.

10-47. What type of single-phase motors cannot be plug-stopped?

10-48. How can we obtain an electrical signal indicating the slip of a synchronous motor?

10-49. What protection is needed for synchronous motors that is not required for a squirrel-cage type?

10-50. When must excitation be applied to get maximum pull-in torque and how can we recognize the appropriate time?

10-51. What is the purpose of a slip frequency relay and how does it work?

10-52. Write out the cycles of operation for the starters in Figures 10-60 and 10-61. Include the following in your explanation:
   (a) What happens on a normal start-up?
   (b) What happens if the dc supply is cut off?
   (c) What happens if the motor loses synchronism due to an overload lasting about 5 s?
   (d) What happens if the motor loses synchronism owing to a long-time overload?

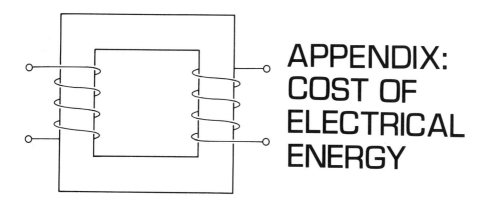

APPENDIX: COST OF ELECTRICAL ENERGY

METERS AND RATE STRUCTURES

Except for loads supplied on a "flat-rate" basis, the use of demand indicating kilowatt-hour meters for commercial and industrial installation is practically universal. These are really two meters in one—a kilowatt-hour (kWh) meter that "counts up" (integrates) the energy used by the load and a (thermal) demand meter that indicates both the present and the peak load that has occurred on the system. Demand meters usually have a (90%) rise time of about 20 min; that is, from a cold start the meter will rise to 90% of the true load value in 20 min, 99% of true load in 40 min, and 99.9% in 1 hour. They therefore are not affected by motor starting currents (which never last more than 1˙min), and in the case of cyclic loads like welders, they indicate the true rms value of the load.

Two varieties of demand meters exist. Some of them respond to the true power (kilowatt) load, while others respond to the apparent power (kVA) load. Usual prac-

tice is to use kilowatt demand meters if the consumer's load is above 90% pf and kVA demand meters if the power factor is less than 90%. In either case the demand meter is reset each time the kWh meter is read, so that the peak load for each month can be established.

The subject of utility rates is very complex, but the technical basis of the rate structure is that the largest single cost of electrical energy is the investment cost of the transmission and distribution equipment. Since the size of the supply equipment (and therefore also its cost) is directly related to the peak kVA load that occurs, it is only reasonable that some charge be levied based upon the kVA demand, as well as a charge for the energy used.

A typical rate structure looks something like this:

Demand charge: $2 per month per kVA peak load demand

Energy charges: $0.07 per kWh for first
(100 × peak load) kWh
$0.04 per kWh for next
(300 × peak load) kWh
$0.025 per kWh for all
over the above

In terms of cost per kWh, such a rate structure penalizes any consumer who operates at a low power factor or at a low number of operating hours per month, and the way to reduce the cost of energy is to reduce either the consumption or peak demand. But consumers have grown accustomed to having unlimited amounts of electrical energy available whenever desired, and some of them are unwilling to forego that freedom, whatever the price.

REDUCING THE COST OF ELECTRICAL ENERGY

Assuming no change in energy consumption, the savings obtained through reduced demand depends on the rate structure and the actual energy consumption. Using the rate structure quoted above and assuming consumption to be over 100 but less than 300 kWh per kVA demand, the savings per kVA reduction of demand are

Savings = $2 + 100 (0.07 − 0.04)

= $2 + $3

= $5 per month per kVA reduction of load

If the energy consumption is over 400 kWh/ kVA demand, the savings become

Savings = $2 + 100(0.07 − 0.04)
+ 300(0.04 − 0.025)

= $2 + $3 + $4.50

= $9.50 per month per kVA reduction of load

There are three ways to reduce peak demand. The first is load shedding or load deferment. If we switch off nonessential loads whenever we approach a peak load condition, or if we can arrange the plant operation so that certain machines will not be operated during peak load hours, we will reduce the demand and therefore the cost of energy per kilowatt hour. There may also be some reduction in the amount of energy used.

The second possibility is to correct the power factor of the plant. Power-factor correction will reduce the kVA demand without affecting kW demand or kWh consumption. If the energy bills are based on kVA demand, power-factor correction will reduce the cost of energy; but if the charges are based on kW demand, power-factor correction will not affect the cost. Some utilities base their charges on "90% of the actual kVA demand or else 100% of the actual kW demand, whichever is greater." In such a case there is no point in correcting the power factor above the 90% level.

The third method is to install equipment that has higher efficiency. For a given plant the peak demand, the energy consumed, and also the monthly energy bill will all be inversely proportional to the efficiency of the electrical machines.

INDEX